The Global Dynamics of Regenerative Medicine

Health, Technology and Society

Series Editors: **Andrew Webster**, University of York, UK and **Sally Wyatt**, Royal Netherlands Academy of Arts and Sciences, The Netherlands

Titles include:

Ellen Balka, Eileen Green and Flis Henwood (*editors*)
GENDER, HEALTH AND INFORMATION TECHNOLOGY IN CONTEXT

Gerard de Vries and Klasien Horstman (*editors*)
GENETICS FROM LABORATORY TO SOCIETY
Societal Learning as an Alternative to Regulation

Alex Faulkner
MEDICAL TECHNOLOGY INTO HEALTHCARE AND SOCIETY
A Sociology of Devices, Innovation and Governance

Herbert Gottweis, Brian Salter and Catherine Waldby
THE GLOBAL POLITICS OF HUMAN EMBRYONIC STEM CELL SCIENCE
Regenerative Medicine in Transition

Roma Harris, Nadine Wathen and Sally Wyatt (*editors*)
CONFIGURING HEALTH CONSUMERS
Health Work and the Imperative of Personal Responsibility

Jessica Mesman
MEDICAL INNOVATION AND UNCERTAINTY IN NEONATOLOGY

Nelly Oudshoorn
TELECARE TECHNOLOGIES AND THE TRANSFORMATION OF HEALTHCARE

Nadine Wathen, Sally Wyatt and Roma Harris (*editors*)
MEDIATING HEALTH INFORMATION
The Go-Betweens in a Changing Socio-Technical Landscape

Andrew Webster (*editor*)
NEW TECHNOLOGIES IN HEALTH CARE
Challenge, Change and Innovation

Andrew Webster (*editor*)
THE GLOBAL DYNAMICS OF REGENERATIVE MEDICINE
A Social Science Critique

Forthcoming titles include:

John Abraham and Courtney Davis
CHALLENGING PHARMACEUTICAL REGULATION
Innovation and Public Health in Europe and the United States

Health, Technology and Society
Series Standing Order ISBN 978–1–4039–9131–7 hardback
(*outside North America only*)

You can receive future titles in this series as they are published by placing a standing order. Please contact your bookseller or, in case of difficulty, write to us at the address below with your name and address, the title of the series and the ISBN quoted above.

Customer Services Department, Macmillan Distribution Ltd, Houndmills, Basingstoke, Hampshire RG21 6XS, England

The Global Dynamics of Regenerative Medicine

A Social Science Critique

Edited by

Andrew Webster
University of York, UK

First published 2013 by
PALGRAVE MACMILLAN

Palgrave Macmillan in the UK is an imprint of Macmillan Publishers Limited,
registered in England, company number 785998, of Houndmills, Basingstoke,
Hampshire RG21 6XS.

Palgrave Macmillan in the US is a division of St Martin's Press LLC,
175 Fifth Avenue, New York, NY 10010.

Palgrave Macmillan is the global academic imprint of the above companies
and has companies and representatives throughout the world.

Palgrave® and Macmillan® are registered trademarks in the United States,
the United Kingdom, Europe and other countries.

ISBN 978-1-349-43924-9   ISBN 978-1-137-02655-2 (eBook)
DOI 10.1057/9781137026552

This book is printed on paper suitable for recycling and made from fully
managed and sustained forest sources. Logging, pulping and manufacturing
processes are expected to conform to the environmental regulations of the
country of origin.

A catalogue record for this book is available from the British Library.

A catalog record for this book is available from the Library of Congress.

10   9   8   7   6   5   4   3   2   1
22 21 20 19 18 17 16 15 14 13

Transferred to Digital Printing in 2013

Contents

List of Figures and Tables

Figures

Tables

Series Editors' Preface

Medicine, health care, and the wider social meaning and management of health are undergoing major changes. In part this reflects developments in science and technology, which enable new forms of diagnosis, treatment, and the delivery of health care. It also reflects changes in the locus of care and burden of responsibility for health. Today, genetics, informatics, imaging and integrative technologies, such as nanotechnology, are redefining our understanding of the body, health, and disease; at the same time, health is no longer simply the domain of conventional medicine, nor the clinic.

More broadly, the social management of health itself is losing its anchorage in collective social relations and shared knowledge and practice, whether at the level of the local community or through state-funded socialised medicine. This individualisation of health is both culturally driven and state sponsored, as the promotion of 'self-care' demonstrates. The very technologies that redefine health are also the means through which this individualisation can occur – through 'e-health', diagnostic tests, and the commodification of restorative tissue, such as stem cells and cloned embryos.

This Series explores these processes *within* and *beyond* the conventional domain of 'the clinic', and asks whether they amount to a qualitative shift in the social ordering and value of medicine and health. Locating technical developments in wider socio-economic and political processes, each text discusses and critiques recent developments within health technologies in specific areas, drawing on a range of analyses provided by the social sciences. Some will have a more theoretical, others a more applied focus, interrogating and contributing towards a health policy. All will draw on recent research conducted by the author(s).

The Health, Technology, and Society Series also looks toward the medium term in anticipating the likely configurations of health in advanced industrial societies and does so comparatively, through exploring the globalisation and the internationalisation of health, health inequalities, and their expression through existing and new social divisions.

This book makes a valuable contribution to the Series by bringing focused and critical attention to 'regenerative medicine', an emerging set of developments in the biosciences, bringing together synthetic biology, embryonic stem cell research, and gene therapy. Such developments may disturb our understandings of the boundaries between body, nature, and identity, and thus require new forms of regulation and governance. Regenerative medicine has attracted interest from the pharmaceutical industry, from clinicians, and from political and other social actors. This volume brings together a group of leading scholars from different disciplines in order to provide a wide-ranging analysis of the meaning and impact of regenerative medicine. In addition, contributors shed light on the ways that the scientific, commercial, and regulatory exigencies combine and converge in order to question the revolutionary claims being made for 'regenerative medicine 2.0'. Contributors provide detailed analyses of what is happening in different parts of the world, including the United States, Europe, and Asia.

Sally Wyatt and Andrew Webster

Acknowledgements

This book is the result of a major research project REMEDiE (Regenerative medicine in Europe) funded by the European Commission's Framework Programme 7 during the period 2008–2011. The project brought together colleagues from a wide range of academic institutions across Europe, reflecting its disciplinary and geographical spread. As Coordinator of the project it was my privilege to work with colleagues who produced work of a high quality and who were always collaborative and open to each other's ideas. We had many very enjoyable meetings, workshops, and conferences in Europe and the United States and engaged with both academic and policy communities as the work developed. The European Parliament as well as national regulatory agencies have engaged with our research results and, via both journal papers and new research funding, colleagues continue to pursue the ideas explored in this book. Updates on our work on regenerative medicine are available on the REMEDiE project page to be found on the Science and Technology Studies Unit's web page at York. I am very grateful to all colleagues for their hard work, collegiality, and good humour. I would also like to thank the Commission for funding the project and for their helpful guidance and advice, particularly Pilar Gonzalez-Pantaleon, Halina Walasek, and Marie-Christine Brichard. Finally, I want to thank Philippa Grand and Andrew James at Palgrave for all their help in seeing the book through to publication.

Andrew Webster

Contributors

Itziar Alkorta is Professor of Civil Law and Bioethics and Vice President for Quality and Innovation at the University of the Basque Country, Bilbao, Spain.

Inigo Miguel Beriain is Senior Research Fellow UPV/EHU in the Inter-university Chair Provincial Government of Biscay in Law and the Human Genome, University of Deusto, and University of the Basque Country, Bilbao, Spain.

Kathrin Braun is currently Dorothea-Erxleben Guest Professor in Political Science at the Otto-von-Guericke University, Magdeburg, Germany.

Haidan Chen is a postdoctoral fellow at Asia Research Institute, National University of Singapore.

Herbert Gottweis is Professor in the Department of Political Science and Director of Life Science Governance Research Platform at the University of Vienna, Austria.

Christian Haddad is a junior researcher at the interdisciplinary Life-Science-Governance research platform and a lecturer in Political Science at the University of Vienna, Austria.

Stuart Hogarth is Research Fellow in the Department of Political Economy at King's College London, UK.

Beth Kewell is Associate Professor at the University of Stavanger's Centre for the Study of Risk and Societal Safety (SEROS), Norway.

Graham Lewis is Research Fellow in the Science and Technology Studies Unit (SATSU) at the University of York, UK.

Michael Morrison is a research fellow at the Centre for Health, Law and Emerging Technologies, Department of Public Health, University of Oxford, UK.

David Rodríguez-Arias is a Juan de la Cierva research fellow at the Philosophy Institute of the Spanish National Research Council (CCHS-CSIC), and belongs to the Group of Excellence G-41 on Applied Ethics at the University of Salamanca, Spain.

Brian Salter is Professor of Politics and Director of Research in the Department of Political Economy at King's College London, UK.

Judit Sándor is a lawyer and Professor at the Central European University, Director of the Center for Ethics and Law in Biomedicine (CELAB) Budapest, Hungary.

Susanne Schultz is a political scientist at the Goethe University Frankfurt/M and working as editor for the NGO Gene-ethics Network in Berlin, Germany.

Marton Varju is a lecturer at the Law School, the University of Hull, and a research associate at CELAB, Central European University, Budapest, Hungary.

Andrew Webster is Professor in the Sociology of Science and Technology, Director of the Science and Technology Studies Unit, and Coordinator of Social Sciences at the University of York, UK.

List of Abbreviations

ATMP	Advanced Therapy Medicinal Products
EMA	European Medicines Agency
FDA	Food and Drug Administration (US)
hESC	human embryonic stem cells
HFEA	Human Fertilisation and Embryology Authority (UK)
iPS	induced pluripotent stem cells
MHRA	Medicine and Healthcare Products Regulatory Agency (UK)
R&D	research and development
SFDA	State Food and Drug Administration (China)
SCNT	somatic cell nuclear transfer
SME	small- to medium-sized enterprises
TE	tissue engineering

1
Introduction: The Boundaries and Mobilities of Regenerative Medicine

Andrew Webster

The analysis of any field of inquiry depends on a clear understanding of where its boundaries lie and what thereby is to be the object of that analysis. However, many scientific fields today are characterised as having highly permeable boundaries, reflecting a number of processes at work – the moves towards transdisciplinarity, the creation of integrative technology platforms (as in the use of bioinformatics or genomics science in diverse domains), and the growth of globalised networks of science seeking to address non-field-specific issues (such as climate change or the stability of ecosystems [Parker et al., 2010]). These processes are evident, for example, in areas such as nanotechnology and informatics. While these dynamics generate considerable activity and churn in an area, there is typically a parallel move towards trying to discipline and police a field, its new entrants, who among those are seen to occupy core compared with more peripheral positions and locations, and so on.

Contemporary developments in the biosciences associated with new biological possibilities – such as seen in synthetic biology, embryonic stem cell research, and gene therapy – also depend on the articulation and integration of different sciences and technologies, such as engineering, physics, and biology within synthetic biology. As a result, the epistemic and professional boundaries of these biosciences are similarly fluid, highly mobile, and yet to be stabilised. While the matter of disciplinary boundaries and their formation is of interest to those within the field of science and technology studies, what has been of especial interest is the ways in which these biosciences not only generate new biological but also social possibilities by disturbing our understanding of the boundaries between the body, nature, and identity (not least in the

1

form of human/animal hybrids), and how these create new 'bio-objects' that are contested, exploited, and require the search for new forms of state and societal regulation and governance (Vermeulen et al., 2012). Indeed, these changes can be so fundamental as to require a radically new approach to the human 'constitution', in both its biological and socio-legal senses (Jasanoff, 2011).

As a field of inquiry, regenerative medicine has attracted major commercial, clinical, political, and popular interest as well as controversy, illustrated by the 2008–2009 debate in the United Kingdom over the licensing of research on so-called 'admixed embryos', the combination of human and animal cells. The contested nature of the area has attracted growing social science interest, with most work focusing principally on human embryonic stem cells (hESC) (e.g., Eriksson and Webster, 2008; Wainwright et al., 2008; Webster and Eriksson, 2008; Gottweis et al., 2009; Ehrich et al., 2010), the emergence of the 'tissue economy' (Waldby and Mitchell, 2006), and the relation between this field and a longer-standing history of managed reproduction and the manipulation of 'life' itself (Franklin and Roberts, 2006). This book builds on this body of work in an attempt to provide as broad-ranging an analysis of the current meaning and impact of regenerative medicine as possible, drawing on a number of different perspectives that provide a critical assessment of the science base and its geographical reach, the parallel hype and promise associated with the field, the uneven and often failed commercial exploitation we have seen, and the political, legal, and bioethical challenges that developments in the regenerative medicine field pose. It does this at a global level, including a detailed analysis of what we are seeing in the United States, Europe, China, and Australia.

In this introductory chapter I provide, first of all, a brief account of the basic characteristics of the field for those less familiar with the area, and then move on to discuss the ways in which different social actors within and outside of science have sought to define and stabilise its meaning and boundaries, and how other processes continually work against this. In doing so I examine how regenerative medicine is mobilised precisely through these contrasting processes and how these different 'mobilities' reflect different purposes and interests that are in tension with each other. I argue that the primary form through which regenerative medicine is developing is through a scientific/corporate/regulatory nexus that is central to the economic and clinical mobilisation of the field. However, this confronts other forms of mobility – found within the corporeal and institutional domains – that pose challenges to the development of the field. I close the chapter by arguing for a 'techno-geography' of regenerative medicine (Oudshoorn, 2011) in order to locate the pathways and spatial configurations of the field that could emerge in the future. The chapter then concludes with an outline of the book.

Establishing the bioscience boundaries of regenerative medicine

There is at present no single, universally agreed definition of regenerative medicine, although recent years have seen a number of attempts in the scientific literature to delineate the field (e.g., Kemp, 2006; Atala, 2007). A commonly adopted view is that it refers to novel biotechnologies that aim to restore, maintain, or enhance tissue, cell, or organ function by stimulating or augmenting the human body's inherent capacity for self-repair (Haseltine, 2001; Daar and Greenwood, 2007; Webster et al., 2011). This differs from traditional drugs or biologicals (such as monoclonal antibodies) in that it seeks not merely to treat and heal the body but to do so by changing the cell structure within the body. Indeed, it is this distinctive aspect of regenerative medicine that has led some (e.g., Mason and Dunnill, 2008) to suggest that it will provide a 'third arm' to medicine, complementing those of conventional drugs and biologicals.

Boundaries of a field are especially difficult to determine where a field is framed by its proponents as 'revolutionary', or as offering a new paradigm for theory and practice, here even likened to the radical change the arrival of Web 2.0 meant for the Internet (Mason, 2007). The two most notable developments in regenerative medicine that might be associated with this language of change were the identification and isolation of hESC, by Jamie Thomson in his lab at the University of Wisconsin–Madison in 1998, and the more recent (2007) creation, by Yamanaka at the University of Kyoto, of 'induced pluripotent stem cells' (iPS) that have the biological potential of embryonic cells without being derived from embryonic tissue as such but from reprogrammed 'adult' (mature, already differentiated) human cells such as skin cells.

The long-term goal of regenerative medicine is to harness the regenerative potential of both hESC and iPS cell lines to restore functionality in the body. Both types of cell lines can, in theory, be used to generate primary tissues and organs through their controlled differentiation in the lab and subsequent transplantation into the human (or animal) body. For example, biomedical scientists are working on restoring heart function through the injection of a specific type of cell line – cardiomyocytes – that has been modified to act as heart muscle. Prior to this *regenerative* paradigm, the field was primarily associated with 'tissue engineering' which, with a focus more on *replacement*, can be traced back some 30 years, involving the use of cells and biomaterial compounds to rebuild damaged body tissue, through skin grafts or bone and cartilage repair, or the use of mature healthy cells transplanted therapeutically to treat haematological (blood) disorders, such as bone marrow cells used to treat childhood leukaemia. A major limitation in the earlier days of tissue engineering was the lack of organs – such as the kidney – available for transplantation, and even when they were available, their use would produce a strong immunological response and the likelihood of rejection by the recipient's body. This matter of the immune response has bedevilled the field and the hope, though yet to be realised, is that hESC and iPS lines may be less immunogenic than normal implants. Even if the immune response can be controlled, a further complication is in preventing cells, once implanted, from becoming carcinogenic (so in the case of iPS cells ensuring the reprogramming is actually secured and stable). On both counts, patient safety is a critical issue. It is important to note that neither hESC- nor iPS-based therapies are yet in clinical use, though a number are undergoing early-stage clinical trials, with large pharmaceutical companies very much adopting a cautious approach to investing in the area (see McKernan et al., 2010).

At present the clinical use of cell therapy techniques and products is in the domain of (non-iPS) 'adult' cell lines, using what are called 'autologous' (i.e., the patient's own) cells. However, while there are some treatments on the market (such as cartilage repair), the cost of developing products, the cost of production, and the size of the markets are all factors, which mean that time to clinical use will take many years for most products currently in development. Successful products and/or procedures are likely to be those that have early links with clinicians who understand precisely what they need and how the delivery system will be able to make new offerings accessible and practicable, in terms of quicker application, greater longevity, and/or enhanced efficacy measurable by clinical end points. Further consideration of this question of product development and associated innovation 'pathways' is provided in much more detail in Chapter 3.

Boundary closures and openings

The account of the field above provides a basic sketch of the bioscientific landscape upon which we can undertake a more sociological exploration of the social, economic, and cultural boundaries of regenerative medicine. In doing so we have to think how boundaries are drawn, contested, consolidated, and negotiated on this landscape through the play of diverse interests. In that sense, there is no natural, or inevitable, boundary to the field. The specific socio-technical landscape of regenerative medicine is populated by cell therapy labs, research networks, regulatory committees, patient advocacy groups, bioclinical collectives (Rabeharisoa and Bourret, 2009), and a range of other heterogeneous elements. These must negotiate not only among themselves but with those actors/agencies found on a wider socio-technical terrain, such as the healthcare system, and the broader regulatory, ethical, and political institutions and cultures that are shaped by but yet constitutive of the field itself.

The stabilisation and thereafter exploitation of any field requires action which in some way defines and marks off its boundary: the 'closure' here is not meant to suggest the closing off or the insulating/isolating of an area but rather establishing the terms on which it claims some degree of warrant, authority, and identity with which those outside the boundaries of the field can and do engage, perhaps better envisaged as a form of 'enclosure'. This involves the policing and disciplining of those working within the field. But this is a process which, paradoxically, affords mobility, scale-up, and globalising effects that can 'act at a distance' (Latour, 1987).

Gieryn's (1983) conception of 'boundary work', initially used to identify the social (rather than any privileged epistemic) bases on which science is demarcated from 'non-science', has been deployed, not least by Gieryn himself (1995) and others (e.g., Bijker et al., 2009) to examine the ways in which scientific boundaries are established more generally, through a mix of individual and collective negotiation, action, and synergy. Gieryn defines boundary work as 'the attribution of selected characteristics to [an institution] (i.e., to its practitioners, methods, stock of knowledge, values and work organization)' (p. 782). As noted above, this involves the disciplining of a field, and this typically is secured across regulatory and innovation networks through standardisation of experimental protocols, data sets, registries of results, shared units of measurement, and agreed criteria for evaluating outcomes (Bowker and Star, 2000). This closure around specific standards in the regenerative medicine field (Eriksson and Webster, 2008 allows what has elsewhere been called the *generification* of technologies (Pollock and Williams, 2009) wherein the alignment of technologies, expectations, and effects is secured beyond an immediate domain and where tacit, localised knowledge is flushed out (Keating et al., 1998). Movement within and beyond the specific regenerative medicine landscape then becomes possible. So, for example, a key issue in the field has been the prolonged attempt to secure agreement over the biological 'markers' or signifiers that indicate cell type, whether, for example, a cell line is simply deemed to be 'research grade' rather than the (more demanding) 'clinical grade', and so on. As agreement over such markers is secured, mobility grows: we might say that the globalisation of markers enables the globalisation of markets.

There are a number of ways in which we can see this stabilisation and policing being undertaken by those within the scientific/corporate/ regulatory nexus. Global stem cell research networks have been established by leading labs including those from Europe, the United States, Japan, and Australia as members of the International Stem Cell Initiative to police the production of stem cell lines via standard operating protocols and the identification of materials to reduce variability across labs, such as the move towards agreeing to a 'defined media' to help the reproducibility of cell batches, remove potential sources of contamination or confounding factors, and enable the comparability of results across labs in different countries. Not only scientists but companies too (those providing the media) have been involved in this process.

A second way in which we have seen boundary closure at work is in regard to the move towards the automation of cell culture/management techniques through scientific equipment manufacturers involved in whole cell bioprocessing. Automation is, as Bartlett (2009) has eloquently suggested, 'the material reification of rationalisation' (p. 75) and in turn helps relieve the tedious work of manual handling of cell batches, their extraction, and the validation of results. It presupposes agreement over how variability in cell lines is to be understood and treated within automated systems, and presumes thereby that what is scaled up has a robust degree of consistency and uniformity. Regulators are very closely involved here since they require this degree of consistency and reproducibility in approving the move towards clinical trials.

A further way in which boundary stabilisation can be secured is through the establishing of what can be called international trading zones through bi- or multilateral stem cell banking agreements that are designed to regularise and agree on (ethical and quality) standards in respect to the procurement of cell tissue, and in the medium to longer term define the criteria through which banks can determine what are 'clinical'-grade lines that would be safe to deploy in clinical trials, and which would be required to meet the terms of the European Tissue and Cells Directives (ETCD) (published between 2004 and 2006). Such agreements (e.g., between the UK and Spanish national stem cell banks) allow for banks to exchange tissue on the assumption that this carries the same socio-technical and regulatory qualities, a form of tissue currency that is seen to be of equivalent material and social value. Reference to the ETCD also points to the way in which the field has seen the growth over the past decade of new forms of regulatory oversight through national and international agencies, notably by the European Medicines Agency and the US-based Food and Drug Administration (FDA). At the same time, the currency of tissue is also expressed in its commodification, which thereby raises questions about the sourcing of tissue, and the political and rights implications of the bioeconomy, a key focus for Chapter 5.

Finally, emerging technologies are often associated with promissory expectations (see Brown et al., 2000), and policy capture can occur as state agencies are caught up in overambitious hype as a result of what has been called 'promissory pressure' (Beynon-Jones and Brown, 2011, p. 640). Recent work by Morrison and Cornips (2012) explores the ways in which news reportage by commercial and trade organisations within the regenerative medicine field has sought to exploit expectations of novelty and disruptive innovation (and so raise venture capital and government interest) while at the same time managing these expectations and endeavouring to frame the field as a stable and reliable site for investment: they show that the passage from the first days of regenerative medicine 'is marked by a transition from early, wild, radical expectations to a more conventional promise that is able to be forced into the "standard" configuration of biotechnology innovation' (p. 19).

While these various ways in which stabilisation of the field has been secured are discussed later in the book there are other processes at work which confound though do not necessarily prevent stabilisation as such. One of these relates to the material nature of the tissue itself – whatever the cell type – which provides the basis for cell lines. As *live* tissue, this is difficult to control not only in the lab experiment but also *in vivo*, whether in animal models or in human trials. The nature of tissue-based biological interactions – cell distribution, engrafting, genetic variability – is seen to vary enormously from tissue to tissue, and patient to patient. Changes over time within the cells, movement of cells *within* the patient's body and in the case of cells derived from other sources *between* human bodies, and seeking to track this and control for it – forms of corporeal and material mobility – pose additional challenges to the bioclinical scientist.

This in turn affects the drive towards the automation and so scale-up of tissue culture. Scale-up in this area is distinct from that of the classic Fordist model, where componentry that can be combined on an individual unit basis can be organised to be mass-produced through a division of labour assembling the *same* standardised components within a factory. Hand-built kits compared with factory-built models are ostensibly the same. In living tissue products there are problems in determining what the components are, how to standardise them, how to optimise them, and how when scaling them up their properties change through their *interaction* with each other as live tissue. Moreover, the variability on living tissue may be worth *retaining* rather than removing as this can be extremely important in the ways in which tissue 'works' within individual patients. So, there is concern that process automation may have a deleterious effect on the functional properties of cells, so affecting their quality and so utility.

In a similar fashion, questions have been raised about whether the conventional protocols used for clinical trials are appropriate to the field (ISSCR, 2008; Webster et al., 2011). The clinical trial, as such, is a bridge between the lab and the clinic, in that sense a key element of so-called 'translational medicine'. There is a significant difference between biological science which is oriented towards discovery and classification compared with medicine which is oriented towards therapy: the former depends on experimentation and replication to establish its claims to truth while the latter depends on reliable know-how and practice-based knowledge to meet its commitment to do no harm. The clinical trial provides a key bridge between these two domains. However, in the case of stem cells and the trialling of cell therapies, the very regenerative potential that makes stem cell treatments appear so promising is also the quality that makes them risky: securing stable implantation can be difficult, cell batches can vary over the course of a trial, end points might be difficult to determine where patients carry a range of co-morbidities, and so on. This issue is explored more fully in Chapter 4.

This matter of variability (or the variability of matter) similarly affects intellectual property (IP) claims within the field. Such claims are only regarded as legitimate, if they are able to describe precisely *what is* being claimed. In emerging fields of inquiry this can be difficult to do since the material and technical nature of the objects that make it up are unstable. This can lead to claims and counterclaims by those seeking to secure IP: for example, the bioscience commercialisation organisation Wisconsin Alumni Research Foundation (WARF), based at the University of Wisconsin, has claimed that the patents it secured (in the United States) on hESC could be extended to iPS cells on the basis that the markers found in hESC appear on iPS lines too. This has not been accepted by the US patent office. The European Patent Office also dismissed WARF's claims to IP on hESC on moral grounds. Even if it were possible to secure agreement on the material specificity of cells, as a more detailed discussion of IP (especially patenting) in Chapters 2 and 7 of this book shows, the very patenting of embryonic cell lines is contested. The recent decision by the European Court of Justice to refuse patenting of hESC reflects how the materiality of the embryo is invested with a special ethical status that precludes its being made a proprietary object.

Such a status produces intriguing regulatory and legal effects when we follow the movement of stem cells across different jurisdictional spaces: this can enable or work against the field. In regard to the former, while embryonic stem cells cannot be produced in Germany or Italy, both German and Italian bioscientists can import them from elsewhere to undertake their research without contravening local regulation. Conversely, patients who have received stem cell treatment elsewhere which is banned in their own country might be deemed to commit a crime on returning home simply by virtue of their carrying stem cell implants, as was argued by some politicians in Minnesota, reflecting local regulatory prohibitions on stem cell research that prevailed between 2009 and 2011. These transjurisdictional mobilities indicate how regenerative tissue is not merely a complex scientific product but is also one that is coproduced by legal and regulatory considerations that shape and impose different sorts of constraints on actors found within the regenerative medicine landscape. These broad issues are discussed in considerable detail in Chapters 6–8.

Those who move across regulatory boundaries in search of treatment that is unavailable in their own country engage in what has been called 'stem cell tourism' as fee-paying patients responding to online adverts from private clinics across the world. There is considerable debate about these treatments in terms of their therapeutic value, implications for patient safety, and the commercial exploitation of the vulnerable (e.g., Lindvall and Hyun, 2009; Levine, 2010; Ryan et al., 2010). While there are clearly ethical and clinical matters of some importance here, the main issue to note at this point is the way in which stem cell tourism is both a challenge to the closure of the field that has to be policed (literally so inasmuch as some US, Dutch, and Japanese clinics have been closed down by legal intervention) and an alternative path to treatment outside the control of the scientific/commercial and regulatory nexus. This in part explains the more recent strategy of the core stem cells organisations – such as the International Society for Stem Cell Research (ISSCR) – to invite patients to report via their web pages on the treatments clinics provide, thereby seeking to monitor and to a degree seeking to influence and report on engagement with the private clinics.

We have suggested above a number of ways in which the consolidation of the boundaries of regenerative medicine is subject to diverse challenges. These express different forms of mobility – within and between bodies, the unstable and variable nature of cells making scale-up, trialling, and IP claims problematic, and the patterns of transjurisdictional mobilities noted above. These can confound the developmental and innovation pathway of regenerative medicine. Indeed, the pathway seems to lead in various directions at the same time. These processes mean that the sorting and stabilising configurations of the field discussed earlier, as Bowker and Star (2000) would note, work most effectively when they are taken for granted, acting in an invisible way to order human action, remaining quite visible and front of stage and very much still in the making.

In many ways, therefore, we might best understand the boundary of the field as being enacted (Mol and Law, 2004) or performed rather than being gradually built up and in some way solidified. Ironically, the claims to its being radical, a 'third' arm of medicine, will only be realised once it becomes de-radicalised and stable in institutional and cultural terms. How this will happen in practice depends on the power of the scientific, corporate, and regulatory nexus to embed regenerative bioscience in healthcare products and practices. How these processes happen at the national, regional, and global levels is the focus of this book.

Structure of the book

The following chapters of the book, written from a variety of disciplinary perspectives within the social sciences, include contributions from international experts in science and technology studies, political science, law, and ethics. Our general integrative principle woven through the chapters is that, as a field in the making, regenerative medicine develops along different innovation paths or 'journeys' (Deuten and

Rip, 2000). These pathways are, for the most part, non-linear, messy, and complex, involving setbacks and detours. It is an especially complex field inasmuch as it combines 'mature' biomedical technologies and therapies (as in tissue engineering and cell therapy that goes back to the 1950s) with emergent and highly unstable science (as in embryonic stem cell research), and as such is characterised by boundary disputes/boundary work and by a regulatory environment which seeks to deploy traditional regulatory oversight of it while acknowledging the limitations of this (both aspects seen, for example, in clinical trial provisions and requirements).

While such disjunctions figure within many advanced medical settings, the globalisation of regenerative medicine, as both promise and practice, has meant that these tensions are ratcheted up at the global level: the rapid growth of 'stem cell tourism' accompanied by moves to constrain this; the growth in China and India of unregulated treatment accompanied by global pressures and preparedness on China's and India's part to introduce new governance processes; the moves towards the international harmonisation of experimental standards in the field of embryonic stem cell research, and regulatory and licensing standards in advanced therapies, accompanied by prevailing localised lab practices and national differences in reimbursement.

Regenerative medicine, we argue, as all fields, is therefore co-constructed and co-evolves across these different fronts, and has a specific geography to it, and patterns of mobility across different boundaries (natural, jurisdictional, and geographical). The book examines the ways in which this creates divergences and convergences, multi-level forms of governance, commercial trajectories that move at different paces and with different 'business models', and the articulation between the demands of the bioeconomy and patenting on the one hand and the normative constraints of ethical and political cultures in regard to the use of oocytes and embryonic tissue for research on the other.

In short, the book aims to clarify the meaning, boundaries, and options/scenarios for future development that regenerative medicine has, the attempts made by different interests (scientific, corporate, political, patient-based) to shape the field and stabilise it so it can be optimised and so mobilised, and the tensions that have been and are involved in attempting to do so. The book moves from a discussion of general trends, innovation pathways, and processes, through to a consideration of biopolitical, regulatory, and ethical issues shaping the field, and concludes with a return to some of the issues sketched out in this introduction, capturing the emergent conclusions we can draw across the chapters and addressing the question whether this field does indeed present a paradigm shift in biomedicine.

The first substantive chapter by Lewis provides a broad map of the global development of the field over the past decade (2003–2012), drawing on data that describes developments in three domains: corporate activity and interests; the sponsorship of clinical trials; and recent patent activity by firms and other organisations. These three domains are important indicators of the strength and the location, or 'geographical shape', of innovation in the field over time.

The corporate data concentrates on developments across the globe, focusing on national and regional trends over time from both historical and thematic perspectives. Substantively, the data relate to a number of primary areas in the field, viz. autologous and allogeneic products, non-stem cell therapies, tissue engineering, and services and related technologies used by public and private labs. Data on clinical trials demonstrates the emerging areas of interest as corporate interests and clinicians seek to move the technology from the laboratory to the clinic and 'real-world' practice. The chapter also provides a picture of recent patent activity, according to therapeutic focus and country/region.

The following chapter by Morrison, Hogarth, and Kewell provides a detailed analysis of European corporate activity within the global regenerative medicine field. It begins by drawing on the primary Science and Technology Studies (STS) literature on expectations and the dynamics of innovation processes to examine both the underlying context of the commercial Regenerative Medicine (RM) endeavour and the competing technological visions and scenarios for the field operating within this paradigm, offering a critical evaluation of these claims. Using empirical data, the authors assess both the commercial performance of European firms, in order to test the fit between industrial visions for the field, and its current patterns of development asking whether issues of scale-up and commercial viability in general make the introduction of new technologies easier as clinical services, and if so why public policy does not attend to this by greater support for what Hopkins (2006) has called the 'hidden innovation system' and closer attention to regulatory issues. Given that European RM has tended to favour autologous, that is, patient-centred, treatments to date, if these cell therapies are likely to be first to the clinic, what are the likely advantages/disadvantages of using an 'IVF-clinic style' approach to delivering cell therapies to patients? And what other business models are being discussed: what, for example, can be learned from the previous history of innovation with bone marrow transplants (Brown et al., 2006; Martin et al., 2006) and from the development of tissue engineering and regenerative medicine in Japan where state-supported clinical innovation has been preferred over private development to a significant extent (e.g., Sleeboom-Faulkner, 2010). This latter part of the chapter also feeds into the discussion of global perspectives in Chapter 8.

Chapter 4, by Gottweis, Haddad, and Chen, explores the regulatory challenges in the field, in particular in regard to clinical trial regulations for stem cell research and development, comparing developments in the United States, Europe, and China. Today, political steering and national government has been gradually replaced by new forms of a 'regulatory state' (Majone, 1999) that is characterised by steering at arm's length – through (semi-)independent regulatory agencies that operate on a regional, national, as well as inter- and transnational scale (Gilardi, 2008). Moreover, state politics must be understood as crucially 'globalised' in the sense that national decision-making is always crucially shaped by transnational politics, business, and the global economy.

The authors explore how, in dealing with pressing regulatory issues at the political and policy level, the United States, Europe, and China face similar challenges. Most visibly, all three systems have undergone more or less significant institutional innovation. Pointing to convergence, all three systems have created and (partially) implemented some kind of risk-based approach to regulating regenerative therapies. By looking at clinical trials with stem cells, the authors show how clinical testing procedures emerge and are co-constructed with different national socio-political configurations.

Securing tissue for stem cell research has depended on a variety of sources, and for many years has been particularly dependent on, the procurement of oocytes, which has triggered an intensive international debate in recent years. A major issue was the question of how the exploitation of women on a global scale might be prevented – especially in light of the health risks involved in oocyte harvesting. Chapter 5 by Schultz and Braun examines the biopolitical economy of oocytes. Based on their empirical research on the infrastructure and logistics of oocyte procurement for stem cell research in Europe, the authors identify a trend towards different forms of commercialisation of oocytes for research in recent years. Against the backdrop of these empirical findings, they undertake a critical review of feminists' debate on oocytes for research, in particular concerning the issue of commercialisation. They argue that three main approaches can be distinguished within this debate, founded on different conceptions of the female subject and her relationship to her body and to the oocyte as bio-object: the female subject as contractual partner, as potential victim, and as regenerative labourer. The chapter suggests the need to transcend the focus on the relationship between oocyte provider and oocyte and to redirect attention to practices of oocyte procurement as social practices embedded within specific bioeconomic rationales and trajectories.

Chapter 6 by Alkorta, Beriain, and Rodríguez-Arias analyses the way European countries have faced the challenge of meeting internationally binding conventions, notably the Oviedo Convention that prohibits the creation of human embryos for research, while still enabling research to be undertaken within their own boundaries. The authors show how modifications in the definition of 'embryo' have made it possible to keep the ban on embryo creation for research purposes while allowing therapeutic cloning. The Oviedo Convention can be presented as a legal 'boundary object' within the field, as it was conceived to construe an ethically valid response to the moral issues thrown up by regenerative medicine across Europe, and yet at the same time has been used in practice as an ethical and legal 'alibi' to circumvent prohibitions on embryo research, mobilising the field across different geographical *and* ethical boundaries.

Chapter 7 by Sándor and Varju examines the boundaries of the bioethical debate concerning the legal protection of biomedical inventions in Europe, though it also introduces elements of comparative legal analysis from other regions (the United States, China, India, and South Korea) for the purpose of examining the global impact of the European debate. The central theme of the chapter is normative multiplicity. It examines how multiplicity has affected the shaping of the limits of obtaining legal protection for (patents on) biomedical inventions consisting of or using human biological material. Normative multiplicity in Europe emerges from multiple perspectives within this landscape. First, the ethical principles and the corresponding regulation of biomedical research and the law on biomedical inventions have developed separately. Second, patent law in Europe is organised on multiple levels and around multiple forums. Third, the applicable bioethical principles, while introducing certain universal benchmarks, acknowledge the pluralism of value judgements in different communities. European value pluralism is an imperative that European-level regulation and the application of the law cannot neglect.

In unveiling this distinct ethical and legal pathway in the discourse on regenerative medicine, the chapter provides an overview of the global, regional, and local ethical and legal framework for biomedical research, discusses the legal responses to the bioethical challenges of patenting human biological material, hESC in particular, examines the challenge of legal and ethical multiplicity in the European setting, and analyses the global impact of the European approach in biomedical research governance.

Over the last decade, 'innovation' has acquired an iconic status in the pantheon of state policies as governments compete for access to the knowledge economies of the future through a search for the appropriate alchemy of innovation governance. Propelled by the imperatives of globalisation, the expectations of their populations and the geopolitics of interstate competition for future economic territories, ambitious governments have uniformly come to regard innovation policy as the key to unlocking the potential offered by the advancement of science. With the advent of the emerging powerful economies, countries such as China, India, and Brazil have aggressively moved to establish their own innovation platforms. In their turn, the developed countries of North America, Europe, and Japan are well aware that they must respond to the challenge posed by the emerging economies to their traditional leadership of scientific innovation.

Nowhere is this dynamic more apparent than in the life sciences and regenerative medicine where the promise of future health, wealth, and happiness forms a staple part of the political narrative. Chapter 8 by Salter examines how the consequent global competition for political advantage in regenerative medicine innovation has intensified the production of new forms of governance designed to enable states and regions such as the European Union (EU) to compete more effectively. Governance has become a knowledge terrain in its own right, fuelled by the political demand that it should constantly reformulate itself to accommodate the requirements of scientific and technological innovation.

The final chapter by Webster reviews and consolidates the principal themes that have emerged throughout the earlier chapters of the book, and returns to one of the key questions that informed this introduction. In reprising the different innovation, regulatory, and scientific pathways discussed in the book as a whole, it asks whether and if so how far regenerative medicine constitutes a paradigmatic shift in the meaning and practice of medicine itself, and how far the relationship between what we have called 'enclosure' and instability is centrally determinant in answering this question.

References

Atala, A. (2007) Engineering tissues, organs and cells, *Journal of Tissue Engineering and Regenerative Medicine*, 1: 83–96.

Bartlett, A. (2009) *Accomplishing Sequencing the Human Genome*, PhD Thesis, CESAGEN, Cardiff University.

Beynon-Jones, S. and N. Brown (2011) Time, timing and narrative at the interface between UK technoscience and policy, *Science and Public Policy*, 38: 639–648.

Bijker, W., Hendriks, R. and R. Bal (2009) *The Paradox of Scientific Authority: The Role of Scientific Advice in Democracies*. Cambridge, MA: MIT Press.

Bowker, G. and S. Leigh Star (2000) *Sorting Things Out: Classification and Its Consequences*. Boston: MIT Press.

Brown, N., A. Kraft and P. Martin (2006) The Promissory Pasts of Blood Stem Cell, *Biosocieties*, 3: 329–348.

Brown, N., B. Rappert and A. Webster (eds) (2000) *Contested Futures: A Sociology of Prospective Techno-Science*. Aldershot: Ashgate.

Daar, A. S. and H. L. Greenwood (2007) A proposed definition of regenerative medicine, *Journal of Tissue Engineering and Regenerative Medicine*, 1(3): 179–184.

Deuten, J. J and A. Rip (2000) Narrative infrastructure in produce creation processes, *Organisation*, 7: 67–91.

Ehrich, K., C. Williams, B. Farsides and R. Scott (2010) Fresh or frozen? Classifying 'spare embryos' for donation to human embryonic stem cell research, *Social Science and Medicine*, 71(12–16): 2204–2211.

Eriksson, L. and A. Webster (2008) Standardising the unknown: practicable pluripotency as doable futures, *Science as Culture*, 17: 57–70.

Franklin, S. and C. Roberts (2006) *Born and Made: An Ethnography of Preimplantation Genetic Diagnosis*. Princeton: Princeton University Press.

Gieryn, T. (1983) Boundary-work and the demarcation of science from non-science: strains and interests in professional ideologies of scientists, *American Sociological Review*, 48(6): 781–795.

Gilardi, F. (2008) Delegation *in the Regulatory State: Independent Regulatory Agencies in Western Europe*. Cheltenham: Edward Elgar.

Gottweis, H. et al. (2009) *The Global Politics of Human Embryonic Stem Cell Science: Regenerative Medicine in Transition*. Basingstoke: Palgrave Macmillan.

ISSRC (International Society for Stem Cell Research) (2008) *Guidelines for the Clinical Translation of Stem Cells*. Illinois: ISSRC.

Haseltine, W. A. (2001) The emergence of regenerative medicine: a new field and a new society, *The Journal of Regenerative Medicine*, 2: 17–23.

Hopkins, M. (2006) The hidden research system: the evolution of cytogenetic testing in the National Health Service, *Science as Culture*, 15: 253–276.

Jasanoff, S. (ed) (2011) *Reframing Rights: Bioconstitutionalism in the Genetic Age*. Cambridge, MA: MIT Press.

Keating, P., C. Limoges and A. Cambrioso (1998) The automated laboratory: the generation and replication of work in molecular genetics, in M. Fortun (ed) *The Practices of Human Genetics*. Boston, MA: Kluwer Academic Publishers, 125–142.

Kemp, P. (2006) History of regenerative medicine: looking backwards to move forwards, *Regenerative Medicine*, 1(5): 653–669.

Latour, B. (1987) *Science in Action*. Milton Keynes: Open University Press.

Levine, A. (2010) Insights from patients' blogs and the need for systematic data on stem cell tourism, *The American Journal of Bioethics*, 10(5): 28–29.

Lindvall, O. and I. Hyun (2009) Medical innovation versus stem cell tourism, *Science*, 324(5935): 1664–1665.

Majone, G. (1999) The Regulatory State and its Legitimacy Problems, *West European Politics*, 22:1–24.

Martin, P. A., C. Coveny, A. Kraft, N. Brown and P. Bath (2006) The commercial development of stem cell technology: lessons from the past, strategies for the future, *Regenerative Medicine*, 1(6): 801–807.

Mason, C. (2007) Regenerative Medicine 2.0, *Regenerative Medicine*, 2 (1) 11–18.

Mason, C. and P. Dunnill (2008) A brief definition of regenerative medicine, *Regenerative Medicine*, 3(1): 1–5.

McKernan, R., J. McNeish and D. Smith (2010) Pharma's developing interest in stem cells, *Cell Stem Cell*, 6: 517–520.

Mol, A. and J. Law (2004) Embodied action, enacted bodies: the example of hypoglycaemia, *Body and Society*, 10(2–3): 43–62.

Morrison, M. and L. Cornips (2012) Exploring the role of dedicated biotechnology news providers in the innovation economy, *Science, Technology and Human Values*, 37(3): 262–285.

Oudshoorn, N. (2011) *Telecare Technologies and the Transformation of Healthcare.* Basingstoke: Palgrave Macmillan.

Pollock, N. and R. Williams (2009) *Software and Organisations.* London: Routledge.

Rabeharisoa, V. and P. Bourret (2009) Staging and weighting evidence in biomedicine, *Social Studies of Science*, 39: 691–715.

Ryan, K. A., A. N. Sanders, D. D. Wang and A. D. Levine (2010) Tracking the rise of stem cell tourism, *Regenerative Medicine*, 5: 27–33.

Sleeboom-Faulkner, M. (2010) Contested embryonic culture in Japan – public discussion, and human embryonic stem cell research in an aging welfare society, *Medical Anthropology*, 29(1): 44–70.

Vermeulen, N., S. Tamminen and A. Webster (eds) (2012) *Bio-Objects: Life in the 21st Century.* London: Ashgate.

Wainwright, S., M Michael and C. Williams (2008) Shifting paradigms? Reflections on regenerative medicine, embryonic stem cells and pharmaceuticals, *Sociology of Health and Illness*, 30(6): 959–974.

Waldby, C. and R. Mitchell (2006) *Tissue Economies: Blood, Organs and Cell Lines in Late Capitalism.* Durham, NC: Duke University Press.

Webster, A. (2011) *Regenerative Medicine in Europe: Emerging Needs and Challenges in a Global Context,* Final Report to the European Commission, SATSU, University of York. Available at www.york.ac.uk/satsu/remedie.reports

Webster, A. and L. Eriksson (2008) Governance-by-standards in the field of stem cells: managing uncertainty in the world of 'basic innovation', *New Genetics and Society*, 27: 99–111.

2
Regenerative Medicine at a Global Level: Current Patterns and Global Trends

Graham Lewis

Introduction

Numerous claims have been made regarding the potential benefits that regenerative medicine will bring to clinical practice. However, robust data on what is happening in the field is difficult to secure. In this chapter we examine that data we have found to have a sufficient degree of accuracy in terms of global trends in three domains: corporate activity and interests; the sponsorship of clinical trials (CTs); and recent patent activity by firms and other organisations such as academic institutions and other 'not-for-profit' bodies. Such data is of interest in its own right because the domains chosen for investigation are important indicators of both the strength and the location (or 'geographical shape') of innovation in this emerging field. In addition, by focusing on these three domains the chapter contextualises the themes discussed and analysed in the chapters that follow.

Among the questions investigated are 'where is commercialisation activity occurring and can we identify national or regional trends over time?' 'What types of product are being developed (e.g., autologous or allogeneic) and in which therapeutic areas are developments taking place?' 'What types of CT are under way, who is sponsoring them, and in which countries are they taking place?' 'Can we see trends in the patents granted with regard to geography, the type of discovery patented, and who is doing the patenting?' Therapies are conventionally provided by the pharmaceutical industry so it is also interesting to ask what stake if any the pharmaceutical industry has in this technology and this aspect of the regenerative medicine 'universe' is also briefly examined.[1]

The significance of corporate data is closely related to CTs activity because a prospective product must proceed through clinical development and obtain regulatory approval – in the case of the European Union (EU) via the Advanced Therapy Medicinal Products (ATMP) Regulation.[2] Unless a company can achieve these steps it is not going to be successful, and the main players and product development to date utilising stem cells are discussed in the section on CTs. Similarly, the involvement or otherwise of major pharmaceutical companies is important because of the well-known difficulties small to medium-sized enterprises (SMEs) experience in translating advanced therapies to the clinic and bringing products to market (Mason and Dunnill, 2008a; Plagnol et al., 2009). As we will see, engagement so far has primarily been through equity investment or direct collaboration with regenerative medicine companies. Such developments, whilst relatively minor at present, particularly in the context of the overall interests of the pharmaceutical industry, are, nonetheless, significant.

Collecting data on regenerative medicine

Assembling accurate data on regenerative medicine is complicated by the difficulty of defining precisely what we mean by the term and then identifying what if anything is new about what is currently occurring in the field. Defining what we mean by regenerative medicine is intellectually interesting but open to dispute. Some commentators have sought to frame it as a new and exciting technology quite distinct from techniques and procedures that 'went before' such as older forms of tissue engineering and associated technologies (Plagnol et al., 2009). Indeed, as pointed out in Chapter 1, Mason, a leading researcher and champion of the technology, goes as far as to describe the field as introducing a new paradigm for medical practice and, in that sense, can be seen as 'revolutionary' (Mason, 2007). However, in clinical terms it is important to remember that the first bone marrow replacement procedure for treating certain cancers, involving treatment with a patient's own cells, was carried out more than 30 years ago (Kemp, 2006; Appelbaum, 2007).[3] Whilst there have been highly significant findings recently that suggest the technology is indeed 'new' and entering a qualitatively different stage in its development, the history of bone marrow transplantation suggests that any definition cannot be based on a simple chronology of events or 'stages' approach to its development.[4]

Also, whilst there have been a number of attempts to describe the world of regenerative medicine, the fast-moving nature of the subject means information quickly becomes dated. A combination of rapid scientific advance plus high levels of speculation and competing claims means it is difficult to obtain a clear picture of the possibilities and likely timeframe for (potential) translation to the clinic. Another solid reason for viewing predictions about the technology's future shape and development as speculative is because neither regulatory pathways nor business models are clearly defined at present (Ginty et al., 2011).[5] As a result there are inevitably questions regarding the field's trajectory and whether the claims made on its behalf are sufficiently robust and therefore likely to occur in the foreseeable future.

This is a particular problem when we consider corporate data as new start-ups appear and other firms merge or disappear altogether over time, and the data shows that these changes can occur over a relatively short period of time. Regenerative medicine has also reached the stage in its development when key clinical studies are commencing aimed at demonstrating the safety or efficacy of potential products or procedures. Or, in a move that can be just as significant in terms of impact, an important trial is abandoned, as recently illustrated by the decision of Geron to pull out of human embryonic stem cell (hESC) development (Boseley, 2011).[6] Similar concerns about out-of-date data ('aging data') apply to measuring patent activity in the context of producing accurate metrics on patent trends. The number of regenerative medicine patents is increasing rapidly and it is reasonable to assume that some may be highly significant and have a major influence on the field whilst others will turn out to be of little value.

Studies in the sociology of expectations (Martin et al., 2008) have shown that statements by key players regarding the future prospects of new biomedical technologies play a crucial role in shaping the technology concerned and this feature has a direct and highly influential bearing on the corporate world associated with it. Descriptions in the literature are often characterised by hyperbole with respect to the future prospects of a technology as firms and researchers seek to promote proprietary platforms and prospective products in order to attract investment for future development (Deuten and Rip, 2000; Brown, 2003; Brown and Michael, 2003). Morrison (2012) has pointed out that 'historiographies can ... be deliberately deployed, in the form of scientific "origin stories", to stabilise and legitimate specific sets of future-orientated claims'. Wainwright et al. (2008) have illustrated this phenomenon in the case of stem cells and the interplay between prior and current expectations about potential applications as discussed by laboratory scientists.

Defining what we mean by regenerative medicine

Before we go any further we need to define what we mean by regenerative medicine. As noted in Chapter 1, from a socio-technical perspective there are no definitive boundaries to the field and it is best described as a complex network of entities (cell therapy laboratories, clinical research networks, regulatory committees, patient advocacy groups, etc.) which, through their interactions with each other and the wider environment in which they operate, serve to constitute and define the field. Nonetheless, in order to collect and analyse quantitative data it is clearly necessary to impose certain boundaries even if these are to an extent arbitrary and open to contestation. Whilst there is no universally accepted definition, broadly speaking, regenerative medicine products are so called because of their ability to bring about, or encourage, the body's inherent capacity for self-repair (Haseltine, 2001; Daar and Greenwood, 2007; Mason and Dunhill, 2008b). Such products differ fundamentally in terms of mode of action from existing therapeutics and surgical interventions in that they offer the prospect of treatments that alter cell structure within the body. This is in contrast to both conventional medicinal products and surgical procedures which work on the principle of either disrupting or masking a cellular process or, as in the case of surgery, direct intervention. As Webster points out in Chapter 1, 'it is this distinctiveness that has led some to suggest that it will provide a "third arm" to medicine, complementing those of conventional drugs and biologicals' (Webster, 2013, p. 3).

In broad terms, our primary areas of interest for data collection are tissue engineering, stem cells, gene therapy, and within these categories, whether interest is directed at autologous or allogeneic products. A fourth and highly significant category is the provision of services such as stem cell technology for conventional drug development in areas like drug toxicity (Neish, 2007; Dutton, 2012). However, from a data collection perspective it is also important to note that many companies which associate themselves with the regenerative medicine label or 'brand' do not necessarily fall within the definition adopted here. Examples include companies developing recombinant growth factors to 'regenerate' tissue damaged by scarring (such as UK-based Renovo); firms employing antibodies to activate and proliferate certain white blood cells in order to combat disease – what can be called induced cell therapy (e.g., NKT Therapeutics, based in Massachusetts, USA); and development of small molecules and therapeutic proteins to induce the regeneration of adult neural stem cells (such as undertaken by the Swedish firm NeuroNova

AB). It is reasonable to exclude such companies because their intended products are essentially either small molecules or macromolecules. Even if these products (or products-in-development) are designed to stimulate cell healing, division, or activation within the body the model of product development is one that essentially follows conventional drug development. Furthermore, these approaches do not involve any significant investment in specialised biomaterials characteristic of the regenerative medicine industry. And importantly, neither are they likely to invoke the same regulatory, financial, or technological barriers which accompany the development of these biomaterials. Essentially, regenerative medicine therapeutics are considered distinct from traditional small or macro-molecule drugs which have a temporally limited intervention – the pharmacological half-life – and are ultimately degraded and discarded by the body. In contrast, while regenerative medicine therapies may not necessarily persist in the body in their applied form, they, nonetheless, have an additive effect, being ultimately incorporated in the body. This basis is also the reason for including gene therapy (but not 'anti-sense' or RNA interference technologies) in the raw data which the chapter relies upon.

Some observers include the so-called 'cancer vaccines' and similar products in their definition of regenerative medicine.[7] Whilst in therapeutic terms this type of agent is increasingly important this category is excluded from the analysis because the focus of such work is on cancer stem cells rather than the manipulation of autologous or allogeneic healthy cells. Having said this, some companies that are not necessarily developing 'regenerative' products or services per se do fall within the scope of our definition. For example, the United Kingdom's Biocompatibles International is, in one sense, a conventional biotech company with a focus on drug discovery. However, it employs a novel drug delivery system comprising adult mesenchymal stem cells genetically engineered to secrete selected therapeutic molecules when implanted in vivo. Since the novel manipulation of stem cells is one of the core characteristics of regenerative medicine, companies like this warrant inclusion in the 'universe' of regenerative medicine as defined here.

The regenerative medicine universe

Methodological issues with corporate data collection

The collection of corporate data from company websites is problematic for all the definitional reasons outlined above, plus, as mentioned already, firms also inevitably seek to promote themselves – what in sociological terms can be called the generation of 'promissory futures' which help shape the trajectory of new medical technologies.[8] As a result claims about the type of work being undertaken and/or its progress or the benefits of a particular platform being developed or utilised by a company may be suspect. It is therefore necessary to carefully scrutinise the claims made by companies and to corroborate them against other sources as best one can.

Conducting web-based research of any description brings a series of other well-documented problems such as variability in the quality of information available, including considerable out-of-date information. The 'language bias' of the web is also likely to skew results, particularly in an area such as regenerative medicine where major effort is being expended in non-English-speaking countries. As reported below, there is considerable activity in Asia but accurate information on the scale of such work may be hard to access via the Internet. Also, information on previous incarnations of now defunct or merged firms can quickly disappear from the web, making it difficult to obtain an accurate picture of trends over time. This type of problem is illustrated by the fact that on occasion it is not possible to determine whether a company is publicly quoted or privately owned using company websites as source or, in some cases, even when a company was founded. Whilst the quality of the corporate data presented here is generally good it is, nonetheless, important to recognise the methodological caveats inherent in such data. Having examined these caveats, the next section provides summative data on corporate activity in regenerative medicine drawing upon the REMEDiE project database (Webster et al., 2011).

Corporate activity in regenerative medicine

The corporate data concentrates on developments across the globe, focusing on national and regional trends over time from both historical and thematic perspectives. The data was collected by comprehensive online searching of company websites over time, with subsequent 'triangulation' using other sources including industry-based newsletters and reports to ensure the accuracy and robustness of the data. The role of SMEs and 'start-up' companies in regenerative medicine is particularly important and data confined to these companies is also presented. The corporate domain is extended to include a number of important academic and other public-sector research centres, particularly those located in the United States (although primary analysis and findings remain focused on global company data). Whilst incomplete in terms of global reach, 'non-corporate' data provides useful information regarding the role of universities and other non-profit institutions in both the progress of the science and the formation of 'spin-out' companies, the latter forming an important part of the field's SME sector. The evidence suggests much of the basic science on stem cells in particular, but also other areas of regenerative medicine is undertaken by academic researchers. Finally, several states now provide some form of government support to encourage R&D and commercial translation, and it is interesting to explore the role of state bodies in efforts to develop advanced therapies and stem cell research, such as the California Institute for Regenerative Medicine (CIRM) in the United States and similar programmes of strategic support in other countries including Germany and the United Kingdom (BIS, 2011; MRC, 2012).[9] Data on the sponsorship of CTs and patent activity later in the chapter helps test this hypothesis.

It is also easy to assume the translation of regenerative medicine will follow the 'pharmaceutical model' of development, with linear development through a series of phases, from pre-clinical animal studies to human trials, with relatively little regulatory involvement in the early stages but considerable intervention in later stages of the process. However, it is quite possible that the picture will be different in the case of regenerative medicine, with 'non-commercial' institutions such as hospitals playing a bigger role in translation to the clinic than typically occurs with conventional medicinal products (Lindvall, O. and I. Hyun (2009). As Whitaker and Foley (2011) argue, in contrast to common perceptions which tend to assume regenerative medicine is mainly about allogeneic medicines and hence much like the pharma model, with products sitting on the pharmacy shelf, clinician-directed interventions involving autologous procedures could well form the bulk of treatments in the short to medium term. According to these researchers, for all manner of reasons – product development, clinical, and investment – for now at least 'the route to the clinic is through the clinic'.[10]

Global developments in regenerative medicine: data on corporate and 'non-profit' organisations

Data on companies covers the seven regions (Europe, North America, Far East, Australia/New Zealand, South America, South Asia, and 'Other') in the REMEDiE project. This data shows that a total of 391 companies are currently engaged in developing regenerative medicine products or services, with a small proportion of these having a product on the market already. Some 34 'Big Pharma' companies (a category that here includes large device, diagnostic and service companies) have a presence in the field and are included in this figure. The number of SMEs active in

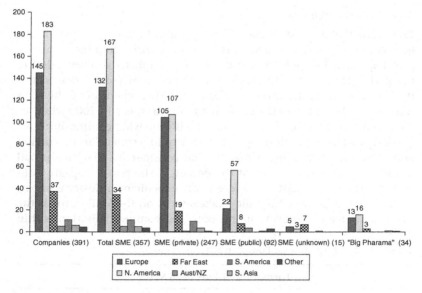

Figure 2.1   RM corporate universe by region in 2010
Source: REMEDiE database

the sector is therefore 357. Of these, 247 firms are private, 92 publicly owned, with the ownership of a further 15 unknown but probably private. In addition, a further 83 'non-profit' organisations were identified (although, as noted above, this data is not comprehensive and the true global figure is much higher), giving a grand total of 473 organisations in the regenerative medicine 'universe'.

Figure 2.1 also shows the regional distribution of companies. Unsurprisingly, North America has the largest number overall with 183 companies, of which 167 are SMEs, with 107 of these privately owned and 57 publicly listed on a stock exchange of some description. Eight of these firms are located in Canada, giving a total of 175 in the United States. Europe has the next largest number with 145 companies, of which 132 are SMEs, with 105 privately owned and 22 publicly traded. According to the REMEDiE database, the Far East has 37 identifiable companies, with 34 SMEs, 19 of which are private and 7 public companies. Global figures for 'Big Pharma' engagement show 17 companies based in North America (of which 3 are J&J subsidiaries or divisions), 14 based in Europe, and 3 located in the Far East (Japan).[11] The form this engagement takes is discussed further below.

Regenerative medicine SMEs in Europe

Table 2.1 (below) lists data on the establishment of regenerative medicine SMEs over time according to type of ownership (i.e., whether privately owned or publicly listed) and the number that closed during the period from 2003 onwards.[12] The figures show considerable activity over a three year period from 2000 to 2002 when 38 of the total number of SMEs in the EU were formed, with a 'spike' in 2000 when 18 companies were established. This finding is somewhat counter-intuitive and may call into question reports of 'lack of investment' in this period and subsequently because it may be that considerable investment had already occurred at the start of the decade.[13] The part European SMEs play in the development of the regenerative medicine industry is of particular interest because EU policymakers focus on the SME sector as the primary source of innovation and potential commercialisation within

Table 2.1   The formation of European SMEs over time

Formation of European SMEs				
Year	No. formed	Private	Public	Unknown[a]	Closed[b]
Prior to 1995	16	12	4	0	0
1995	0	0	0	0	0
1996	4	3	1	0	0
1997	10	6	3	1	0
1998	5	3	2	0	0
1999	4	3	1	0	0
2000	18	13	5	0	0
2001	9	6	1	2	0
2002	11	10	1	0	0
2003	4	2	2	0	1
2004	2	1	0	1	0
2005	4	4	0	0	1
2006	7	7	0	0	2
2007	7	6	0	1	2
2008	4	4	0	0	3
2009	3	3	0	0	6
2010	5	5	0	0	2
Subtotal	113	88	20	5	17
No data[c]	19				0
Total SMEs[d]	132				

[a]Unknown corporate structure, most likely private.
[b]No data on closures prior to 2003; in March 2011, $n = 2$.
[c]No data on foundation date.
[d]Total number of SMEs recorded March 2011.

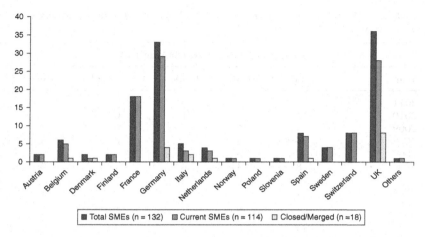

Figure 2.2 EU SMEs by member state (plus Switzerland and Norway) 2003– 2010

regenerative medicine and within the biomedical science more generally (Hogarth and Salter, 2010).

Figure 2.2 (above) shows the distribution of SMEs by EU member states with regard to both total number of SMEs in existence over the period 2003 to early 2011 ($n = 132$) and current SMEs ($n = 114$). These figures show the major national players, which defined by the number of companies located in a particular member state are the United Kingdom, Germany, and France. It also demonstrates the efforts undertaken by Spain ($n = 8$) and Switzerland ($n = 8$) to build a presence in the industry. However, given this data is based on a somewhat crude measure in terms of innovatory capacity (i.e., the number of companies in each EU member state), it tells us nothing about the merits or otherwise of the technologies being exploited by the companies concerned. Nor does it offer guidance with regard to the likelihood these companies will be successful in terms of eventual commercialisation. This caveat is illustrated by the example of Spain which, as noted above, is fifth overall in number of firms yet has one of the most successful companies in Europe, Cellerix – though the company became a subsidiary of Belgium firm Tigenix in early 2011.[14]

US corporate universe

As noted above, the largest concentration of corporate activity is located in the United States – a finding that is no surprise given the size of the

Table 2.2   The formation of SMEs engaged in developing products and services in North America

	Formation of SMEs in North America[a]				
Year	Total	Private	Public	Unknown[b]	Closed[c]
2011	2	2	0	0	2
2010	1	0	1	0	3
2009	4	1	3	0	2
2008	6	4	2	0	3
2007	9	8	1	0	1
2006	8	3	4	1	0
2005	12	7	3	2	0
2004	12	9	3	0	1
2003	7	5	2	0	0
2002	10	8	2	0	2
2001	12	9	3	0	0
2000	9	6	3	0	0
1999	6	3	3	0	0
1998	7	6	1	0	0
1997	6	5	1	0	0
1996	4	2	2	0	0
1995	7	7	0	0	0
Prior to 1995	35	15	20	0	0
Subtotal	157	100	54	3	14
No data[d]	10	7	3		1
Total SMEs[e,f]	167	107	57	3	15

[a]Includes Canada $n = 9$ (2 closures in total; Vario acquired by Fate (US), 2010; BioSyntech acquired by Piramel (India), 2010 after bankruptcy).
[b]Unknown corporate structure, most likely private.
[c]Breakdown of closures: public $n = 6$; private $n = 9$.
[d]No data on foundation/closing date.
[e]Total number of SMEs on WP7 database.
[f]Data not collected prior to 2002.

US biotech industry and the availability of capital markets to foster start-ups. Table 2.2 provides data on the formation of US and Canadian SMEs engaged in developing regenerative medicine products and services. Like the industry in Europe, the US industry is dominated by SMEs.

Distribution by US state

The distribution of companies by US states is also useful as it identi-fies 'hot spots' of activity, particularly with regard to commercialisation and 'bringing to market'. Looked at in terms of the biotech sector as a whole, it is no surprise that states most active in other bioscience and

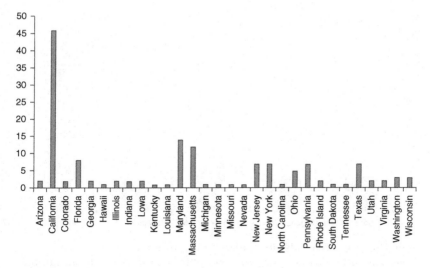

Figure 2.3 US corporate universe by state – current SMEs

health care sectors are also the most active in regenerative medicine. However, as in the case of European SMEs, the data says nothing about the relative importance of individual companies or their location, in efforts to develop advanced therapies. What it does illustrate is the regional infrastructure (and hence networks) states have built around regenerative medicine, often with state funding. Such efforts often reflect wider efforts to develop the biosciences and innovatory capacity more generally and are usually coupled (and often co-located) with major university and other research centres in the state concerned (Figure 2.3).[15]

Regenerative medicine development by product type

The types of activity that firms are engaged in, as well as their location, are also important to consider. Firms engage with the field in different and often complex ways and numerous types of approach to potential treatments and interventions are evident. In this section we examine this engagement according to four categories: autologous, allogeneic, 'other' (mainly gene therapy), and service provision. Figure 2.4 (below) shows a breakdown of the different approaches adopted by companies active in the field. Based on a global total of 402 companies, there are slightly more developing autologous treatments or procedures (*n* = 80) compared with the number of companies developing allogeneic

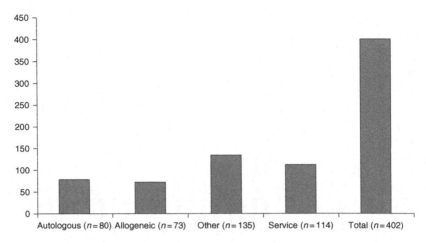

Figure 2.4   The regenerative medicine 'universe' by product type
Note: The difference between total number of entries and total number of organisations reflects the fact that some 46 firms claim to be engaged in more than one area of work.

products ($n = 73$). However, these two categories are outnumbered by companies engaged in what are termed 'other' activities, which in the main comprise the development of treatments involving gene therapy approaches ($n = 133$).[16] The figure of $n = 133$ contrasts with a 2010 report which puts the number higher, with 189 companies with gene therapy programmes and 354 US studies being conducted as 'developers continue to pursue game-changing treatments for some of the most difficult-to-treat diseases' (Martino, 2010).

A total of 114 companies are found to be engaged in service activities, a category that includes the development of diagnostics for drug development using techniques based on induced pluripotent stem cells (iPS), and provision of cell manufacturing services. Whilst the former does not result in regenerative medicine 'products' as such it is an increasingly important area in terms of value and development timescale (i.e., 'time to market'). As noted previously, providing crude data like this does not identify the leading players in the field nor say anything about the probability of either autologous or allogeneic products arriving in the clinic, nor the likely timescales relating to such processes. The figures do, however, provide a picture of the relative effort being expended across these four categories.

Companies engaged in developing regenerative medicine services

It is worth providing some examples of the range of services, as opposed to products, that are being developed. In 2010 Axiogenesis AG and iPS Academia Japan Inc. announced an agreement by which the former obtained access to the extensive portfolio of iPS cell technology pioneered by Yamanaka in Kyoto (Yamanaka, 2009), to produce iPS-derived cardiomyocytes and other cell types. Reprocell and Life Technologies are two other companies which have also licensed the same technology for drug discovery applications including toxicity assays and ADME (absorption, distribution, metabolism, and excretion of the drug) testing in the belief that iPS cell technology will become a mainstream technology in drug screening in the near future. In the same year, Becton Dickenson (BD), a global medical technology company with a focus on cell reagents, entered into an agreement with Fate Therapeutics for joint development and commercialisation of iPS tools and technologies for drug discovery and development, and, in addition, manufacture of cell-based therapies[17] using Fate's iPS platform and BD's commercial expertise to market such products on a worldwide basis. Ecytcell, a wholly owned subsidiary of French firm Cellectis, is a European biotech currently developing iPS cells for human therapeutic and research tool applications. As noted above, iPS developments directed at toxicity screening and drug development are likely to be the most significant in the short to medium term. However, Celgene Cellular Therapeutics (formerly Anthrogenesis), a company that pioneered the recovery of stem cells from human placental tissue, is reportedly developing proprietary technology for collecting, processing, and storing placental stem cells for therapeutic applications in cancer, as well as autoimmune, cardiovascular, neurological, and degenerative diseases. Finally, examples of contract manufacturing organisations (CMOs) include Lonza, the largest for cell-based therapies, which can now produce 100 billion cells per batch which allows for 50–100 million cells per dose; and Angel Biotechnology in the United Kingdom, which is licensed by the UK Medicine and Healthcare products Agency (MHRA) under EU CT Directive (2001/20/EU) to manufacture stem cells for CT use.[18]

CTs and regenerative medicine

Mapping CTs activity provides another measure of the extent to which translation to the clinic is occurring in an emerging field like regenerative medicine – who is sponsoring trials, where they are located (which is not necessarily the same country as the sponsor), what type of cell therapy (autologous or allogeneic), and at what stage (i.e., Phase I, II, or III). In short, a snapshot of trials demonstrates the areas of therapeutic importance within the field as corporate investors, researchers, and clinicians seek to move the technology from laboratory to clinic and 'real world' practice. Tracing developments in clinical studies can also inform analysis of emerging regulatory frameworks. This is an important aspect because the expectation is that products will require a very different regulatory framework compared to conventional medicinal products.

The information discussed in this section relates to the various regions of the world, plus national data for six key countries for illustrative purposes (the United States, the United Kingdom, China, Japan, South Korea, and India). These countries are selected because they are amongst the most active in terms of trials as well as providing a global 'spread' across the regions. Comparisons between countries and regions provided useful information about the positioning of European companies relative to global competitors (e.g., type of product, translational processes, the capacity to conduct CTs, regulatory structures) in addition to trends and overall prospects in this fast-moving field. Raw data was obtained from the publicly available US National Institutes for Health (NIH) CTs database (clinicaltrials.gov), and consists of cumulative data for the period 2003–2010 using the search terms 'autologous cell therapy', 'allogeneic cell therapy', and 'stem cell(s)'.

As with corporate data, it is worth noting that collection of CTs information is not a straightforward exercise. One reason is the lack of compulsory registration in some countries with the result that many trials are not registered with one of the several registries now in existence. There is also an absence of clear and enforceable regulatory guidelines on what constitutes a CT in parts of the world. Even where regulation is in place there may be different interpretations of what constitutes a cell therapy.[19] And the use of unproven therapies in some countries means the distinction between CT and clinical practice is effectively removed in such instances. This is illustrated by recent disclosures that the Chinese authorities are having difficulty enforcing new rules covering stem cell treatments (Cyranoski, 2012a); and that the Texas Medical Board is allowing a local company, Celltex, to use expanded adipose stem cells (eASCs) for treatments that are both unproven and without Food and Drug Administration (FDA) approval (Cyranoski, 2012b).

Methodological questions raised by the use of clinical trials.gov database

The use of the US NIH's clinicaltrials.gov database raises a number of methodological questions, many of which match those raised in earlier sections: to what extent does data obtained from this source accurately reflect global activity given that the database is operated by a US government agency? Do all countries routinely submit information on trials? It seems unlikely this is the case, although there are a number of Chinese trials recorded. It is also possible that data submission rates will increase over time. The NIH database also generates different results depending on whether information on the trial phase is included in the search compared to if these variables are omitted. Whilst the variation is small it is, nonetheless, disconcerting and creates concerns about the robustness of the content in public databases like clinicaltrials.gov particularly in the context of regenerative medicine.[20]

More generally, as noted above, there is also a lack of clarity regarding what constitutes a CT and the boundary between trial and therapy is not as well-defined as in other therapeutic areas. For example, hospitals in some countries offer stem cell or other procedures as treatments although they have not been the subject of formal CT protocols. In part this is because they have been progressed through the 'hospital exemption' route[21] but also because of different governance arrangements. On a more subtle level, development paths for autologous treatments in particular are characterised by clinician-led procedures rather than 'off the shelf' products and this non-linear process may blur the boundary between a CT and treatment. Finally, we should note that the European Medicines Agency (EMA) has recently made its 'EudraCT' database publicly available (https://eudract.ema.europa.eu/) and this will provide another data source for this type of information in the future.

What does CTs data tell us about the field of regenerative medicine?

In the context of Europe, CTs data collected via industry-orientated sources as part of the corporate universe work discussed above shows no automatic correlation between the number of companies in EU member states and CTs activity. For example, figures for some of the key EU players are given in Table 2.3.

This finding is perhaps not surprising because of the total German SMEs ($n = 37$), 14 have products on the market already (and hence studies on these products are likely to have been completed) plus many

Table 2.3   Comparison of number of SMEs in EU member states and clinical trial activity

EU member state	No. of SMEs	No. of clinical trials
Germany	37	6
UK	36	9
France	19	2

of these firms are in the older, more-established tissue engineering sector. In other words, whilst the number of SMEs does not necessarily provide an indication of the relative success or otherwise of a country's regenerative medicine industry in terms of commercialisation, the figures do tally with the fact that Germany has a well-established tissue engineering sector within the industry. In the case of France, inspection shows that several of the companies are at an early stage in the product development cycle and R&D is at the pre-clinical stage in all examples. The United Kingdom, with a similar number of firms to Germany, appears between Germany and France with regard to the number of trials, with more trials than Germany reflecting the fact that the United Kingdom has less involvement with older tissue engineering products and greater investment in more recent developments – in other words, the sector in the United Kingdom is arguably more advanced in terms of product development ('products-to-be').

CTs involving 'autologous cell therapy', 'allogeneic cell therapy', and 'stem cells'

For illustrative purposes, data is provided on six countries (plus 'all countries') over time: the United States; the United Kingdom; China; Japan; South Korea; and India, for autologous and allogeneic cell therapy and stem cells. The 'X' axis displays the number of CTs.

The United States conducts more CTs than any other country in autologous cell therapy.

However, as Figure 2.5 shows, US dominance has decreased in recent years as a proportion of all trials conducted involving this cell type (Figures 2.6 and 2.7).

Details on the main global players in the stem cell field, with CT status of potential products, are presented in Table 2.4, with a brief commentary on notable developments.

Among the recent developments of note because they utilise embryonic stem cells, and because they mark the first of such trials,

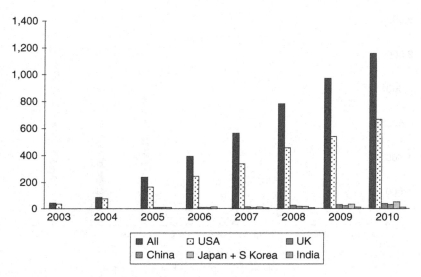

Figure 2.5 Clinical trials involving 'autologous cell therapy' over time
Source: www.clinicaltrials.gov

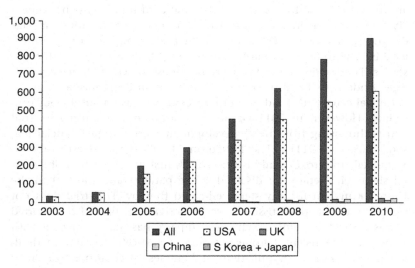

Figure 2.6 Clinical trials involving 'allogeneic cell therapy' over time
Source: www.clinicaltrials.gov

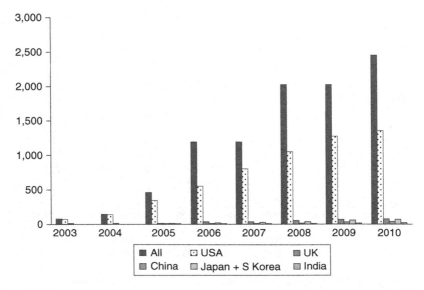

Figure 2.7   Clinical trials involving 'stem cells' over time
Source: www.clinicaltrials.gov

are the Geron trial for treatment of spinal cord injury, which received
US FDA approval in 2010; and the ACT trial for Stargardt's macular
dystrophy, also with FDA approval. In both examples, the platform
used is in vitro fertilised blastocysts (derived from embryos) as the cell
source. The London Project to Cure Blindness, based at University College London (UCL), also has a CT under way utilising retinal pigment
epithelial cells, with funding from Pfizer, as well as an autologous procedure. However, in 2011 Geron announced it was halting its trial
and withdrawing from hESC development work, citing financial reasons (Boseley, 2011).[22] The ReNeuron PISCES (Pilot Investigation of
Stem Cells in Stroke) study is the world's first approved trial of a neural stem cell therapy for disabled stroke patients and the first in the
United Kingdom for any stem cell-based therapy. This trial does not
use embryonic-derived tissue however, deriving its tissue from aborted
foetus instead, a point which the company has made much of with
respect to the recent European Court of Justice's decision to disal-
low patenting on embryonic-derived therapies (Blackburn-Starza, 2011;
European Court of Justice 2011b).[23]

As Table 2.4 below shows, of the 16 main companies or projects worldwide currently developing therapies, more than half have competing

Table 2.4 Principal companies/projects in the stem cell field with clinical trials programmes

Company	Cell type	Cell source	Type	Phase	Indication
Aastrom Biosciences	Non-ESC	Bone marrow	Autologous	Phase III/Phase IIb (planned)	CLI/dilated cardiomyopathy
Advanced Cell Technology	ESC	In vitro fertilised blastocysts ($\rightarrow$ retinal pigment epithelial cells)	Allogeneic	Phase I/II	Stargardt's macular dystrophy/AMD
Athersys	Non-ESC	Bone marrow	Allogeneic	Phase I/II	Acute MI & GvHD/stroke
Azellon Cell Therapeutics	Non-ESC	Adult stem cells from iliac crest (in bioscaffold)	Autologous	Phase I/IIa	Meniscus repair ('Cell Bandage')
BioHeart	Non-ESC	Thigh muscle	Autologous	Phase II/III	Congestive heart failure
BrainStorm	Non-ESC	Bone marrow	Autologous	Phase I/II	ALS
Celgene	Non-ESC	Placenta	Allogeneic	Phase I/II	Crohn's disease/MS/RA
Cytomedix (formerly Aldagen)	Non-ESC	Bone marrow	Autologous	Phase I/II	CLI/stroke
Cytori Therapeutics	Non-ESC	Liposuction	Autologous	Phase I	AMI and chronic myocardial ischemia
Geron[a]	ESC	In vitro fertilised blastocysts	Allogeneic	Phase I	Spinal cord injury

Table 2.4 (continued)

Company	Cell type	Cell source	Type	Phase	Indication
International Stem Cell	Non-ESC	Unfertilised parthenogenetically activated oocytes	Allogeneic	Preclinical	AMD
London Project to Cure Blindness[b]	ESC	In vitro fertilised blastocysts (→ retinal pigment epithelial cells)	Allogeneic/ Autologous	Phase I/II	AMD
NeuralStem	Non-ESC	Spinal cord of eight-week foetus	Allogeneic	Phase I	ALS
Osiris Therapeutics	Non-ESC	Bone marrow	Allogeneic	Phase III/II	GvHD, Crohn's/T1D, AMI
Pluristem Therapeutics	Non-ESC	Placenta	Allogeneic	Phase II (planned)	CLI
ReNeuron	Non-ESC	Neural SCs with expansion	Allogeneic	Phase I/II (planned)	Ischaemic stroke
Stem Cells	Non-ESC	Foetus	Allogeneic	Phase I/II	Spinal cord injury

[a]Geron to complete Phase I study only – see text.
[b]UCL Institute of Ophthalmology, Moorfields Eye Hospital, and University of Sheffield Centre for Stem Cell Biology, UK. AMD, age-related macular degeneration; CLI, critical limb ischemia; ALS, amyotrophic lateral sclerosis (also known as Lou Gehrig's disease); GvHD, graft versus host disease; AMI, acute myocardial infarction; T1D, type 1 diabetes.

programmes in three major disease areas: cardiovascular, gastrointestinal, and the central nervous system (CNS). Within cardiovascular, two areas of disease are focused on: critical limb ischemia (CLI)[24] and acute myocardial infarction (AMI).[25]

Currently there are three major firms with clinical programmes to treat CLI, all US-based: Aastrom Biosciences, who are developing an autologous procedure and, with a Phase III trial under way, thought to be the most advanced clinical programme involving stem cells; Cytomedix (formerly Aldagen) (ALD, 301); and Pluristem Therapeutics, who are developing PLX-PAD. The main companies for AMI ($n = 3$) are Osiris Therapeutics (Prochymal), Atherysys (Multistem), and Cytori Therapeutics (Celution System). Of note is the marketing authorisation granted by Health Canada for Osiris' Prochymal, derived from bone marrow, for the treatment of graft versus host disease as this marks the world's first 'off-the-shelf' stem cell product.[26] As noted later, both Atherysys and Cytori have recently received investment from pharmaceutical companies. Interestingly, Osiris is continuing Prochymal development despite the termination of its collaboration with Genzyme by the latter's new owner, Sanofi.

Mention should also be made of European companies conducting non-stem cell development programmes. For example, Belgium's TiGenix has EU marketing approval under the ATMP Regulation for ChondroCelect, a non-stem cell cartilage repair product. However, in the United States the FDA has demanded another trial before the company can submit the product for US approval, meaning a five-year delay before US approval is likely. The company has also announced CTs of the company's expanded allogeneic adult stem cell product, C × 601, derived from adipose tissue, for Crohn's disease (Phase III) and rheumatoid arthritis (Phase II). TiGenix also has a CE-marked approval for a bioscaffold, TGX002, for aiding joint repair, which is close to entering the market. Spain's Cellerix recently conducted Phase III trials on an autologous treatment, C × 401/Ontaril, and on a second product, C × 501, currently in Phase II. These programmes have reportedly received setbacks and their future is unclear at this time. All these products are designed for treatment of fistulas and skin regeneration. But what may turn out to be as important is the 2011 merger of TiGenix and Cellerix to form Europe's largest and most successful regenerative medicine company to date. Finally, the UK-based Intercytex, once a leading European SME in regenerative medicine, developed a series of non-stem cell autologous and allogeneic cell therapies for wound care, facial rejuvenation, and hair loss, all of which had been under clinical development for a number of years. Typical of the company's portfolio were ICX-SKN and Cyzact (formerly ICX-PRO), topical wound care products designed to stimulate active repair and closure in persistent chronic wounds, with Cyzact completing a Phase III trial. However, in a reflection of the difficulties companies face bringing products to market, this and much of the company's other intellectual property (IP) was recently sold to other parties, although the company has retained rights to one product, Valveta, which is continuing in clinical development.

Patent data on regenerative medicine

Another important way of mapping developments in an emerging biotechnology is to examine the allocation and distribution of IP relating to that technology. Patent data so interpreted can act as an 'index' of invention and innovation (Nelson, 1998; Suarez-Villa, 2000) and in this way provides a means of assessing both the growth of a specific biotech sector such as regenerative medicine (Reiss and Strauss, 1998; Bergman and Graff, 2007b) and the process of 'turning science into business' in the bioeconomy (Oldham and Cutter, 2006). This section provides a picture of recent patent activity according to type of organisation, therapeutic focus, and country/region for a subset of RM, namely, claims granted that refer to stem cells for the period November 2008–June 2010.

Why collect data on patent activity in regenerative medicine?

The collection of data on patents is methodologically difficult for all but experts in patent procedures and patent law. Part of this difficulty arises from the definitional problems discussed earlier in the chapter. What exactly we mean by the term and how and what parameters we use when we set about collecting data will influence the nature and meaning of the information collected. In this section (i.e., for patents alone) the definition used and subsequent analysis is restricted to 'stem cell patents' as defined by the UK Intellectual Property Office (UKIPO).[27]

Patents protect IP and give the patent owner(s) the right to capitalise on this property. As noted above, patents are seen by policymakers as a key indicator of trends in innovation in a particular field and hence prospects for future economic growth. Such trends are often analysed in the context of national innovation strategies and viewed as indicators of national (or regional) strengths relative to other countries (or regions) in domains like the biosciences. In this way, measurement of patent activity is commonly used as an indication of the success or otherwise of a particular innovation strategy.

Other metrics used in studying the location of innovation have tracked factors such as the flow of scientists between countries and the output of scientific literature (Friedman, 2010). Although metrics such as these have been used to assess basic science output in areas such as pharmaceuticals, they lack precision since the productivity of individual scientists and the quality and impact of research papers are generally not addressed. On the other hand, patents can be linked (eventually) to tangible outputs and this allows one to calculate a measure of productivity in different regions. However, potential flaws in measurement remain, such as variation in the inventor criteria. For example, the US Patent and Trademark Office (PTO) inventor criteria are well-defined: the inventor must contribute to the conception (and not merely the reduction to practice) of the invention and must maintain intellectual domination of the work (US Department of Commerce, 2010).[28] But this may not be the case in other jurisdictions. There may also be variation in the way decisions are made in different jurisdictions. When examining pharmaceutical innovation, Freidman relied on US PTO data to avoid such anomalies because companies must list the patents protecting their products with the US FDA (in the so-called FDA Orange Book), including all inventors. This linkage between drugs, patents, and inventors allows a more robust assessment of the location of innovation. Such an approach is not possible with regenerative medicine because there are few approved products at present. But patent agencies such as the UKIPO and the European Patent Office (EPO) have developed sophisticated coding systems that describe and differentiate between different types of research activity within the field. With the UKIPO data on stem cells used here, four categories are defined: embryonic, adult, iPS cells, and 'other'.

Earlier in the chapter we examined the location of firms as a measure of innovation or, more accurately, as a measure of emerging trends in national and regional innovation. This type of approach is valid in the case of SMEs but in the case of larger companies the method may be less robust due to firms having more than one location. The global HQ is not necessarily the national or global location where the company's research is conducted (Light, 2009). Also, regenerative medicine forms only a very small part of total R&D conducted by large pharmaceutical companies. However, with regenerative medicine SME location and R&D location are likely to be the same and hence company location is likely to be a more accurate measure of trends than is the case with more developed technologies.[29]

What data is presented?

It is important to note that patents are often granted to a combination of organisation(s) and individual(s), and, on occasion, jointly to a company and academic institution and/or individual(s). Although it is difficult to assess the significance of such joint arrangements, the number of patents granted which fall into such categories is listed for completeness (see Table 2.5). There is also the issue of what is meant by the type of organisation ('Academic etc.' and 'Company'). The category 'Academic etc.' refers to academic institutions, government laboratories, and other non-profit organisations and is similar to that used above for corporate and 'non-profit' bodies. Although difficult to define precisely, differentiating between patent activity undertaken by the corporate sector and academic and non-profit institutions provides a useful indicator of the importance of the two sectors with regard to patenting and the importance the sectors themselves attach to such activity.

What does the data show?

Perhaps the most striking finding is the overwhelming prevalence of patents granted to 'academic and non-profit' institutions compared to commercial organisations. This finding provides little immediate information about the potential value of any individual patent, in terms of

Table 2.5 Patents granted according to type of assignee (November 2008–June 2010)

Type of assignee	Patents granted
Total no. of patents recorded	317
Company	148
Academic etc. (incl. joint)	159
Academic etc. only	138
Individual(s) assignee only	23
Company and Individual(s) AND Academic etc.	1
Company and Individual(s)	4
Academic etc. and Company	5
Academic etc. and Individual(s)	6

either clinical worth or commercial advantage which may accrue from ownership. Nonetheless, it does demonstrate the part played by universities and other 'non-profit' bodies in stem cell research (Hopkins et al., 2006). The figures also provide an interesting contrast to data presented earlier in the chapter which tends to emphasise the role of the corporate sector in the wider regenerative medicine universe. Overall, the data illustrates the key role that universities and other academic-type institutions play in regenerative medicine, and the biosciences more generally, in terms of patent activity compared to corporations, which rely on investment from profit-seeking individuals, venture capital, and other investment sources (Table 2.6).

Data on the number of patents by country for the period November 2008–June 2010 show that the United States is the most active in patenting stem cell research with 171 patents granted over the period. In Europe, the leading countries are Germany (18), the United Kingdom (12), Switzerland (7), France (6), and Italy (5). None of these totals are surprising apart perhaps from the omission of Spain (1) from the list. Spain is interesting because it has been a leading European player in terms of attempts to commercialise the technology, yet the data shows only one stem cell patent granted over the period examined. The other state with a relatively large number of stem cell patents granted is Israel, which reflects the country's significant science base and well-developed biotechnology industry.

Turning to Asia, the most active states are Japan (28), Korea (13), Taiwan (7), and Singapore (5). The number of patents granted to institutions in China (2) appears low. This probably reflects both the late entry into patenting protocols by China and perhaps a relative lack of interest in patent protection historically compared to other countries. Given reports on China's interest in the field, it most certainly reflects an 'under-reporting' of China's global position in stem cell research (Figure 2.8).

Data according to type of patent granted

Using data supplied by UKIPO, there were a total of 314 stem cell patents granted over the period November 2008–June 2010. Of this total, 215 were classified as involving 'adult' stem cells, 67 were referred to as embryonic stem cells, and 35 were classified as 'embryonic/pluripotent' patents. There were ten entries recorded as 'other types' of stem cell manipulation, according to UKIPO records. Details are provided in chart form in Figure 2.9.

Table 2.6   Stem cell patents by assignee country and by patent office (November 2008–June 2010)

Assignee country	No. of patents (n = 314)	Notes	Patent office		
			EPO (n = 84)	USPO (n = 210)	UK IPO (n = 20)
Australia	4		1	3	
Canada	12		3	9	
China	2	Including 1 with Taiwan	1[a]	1	
Cyprus	1		1	0	
France	6		4	2	
Germany	18	Including 1 with USA	13[b]	5	1[c]
India	1		0	1	
Israel	11		3	6	2
Italy	5		3	2	
Japan	28		14	15[d]	1
Korea	13		6	6	
Netherlands	1		0	1	
Singapore	5		0	2	3
Spain	1		0	1	
Sweden	4		1	1	2
Switzerland	7		5	2	
Taiwan	7	+1 with China – see above	1	6	
UK	12	+1 with USA – see below	3[e]	6	4
USA	171	+1 with Germany; 2 with Japan; 1 with UK – see above	26[f]	140	7

[a] Jointly with Taiwan assignee.
[b] Includes 1 jointly with US assignee.
[c] Jointly with US assignee.
[d] Includes 2 with US assignee.
[e] Includes 1 jointly with US assignee.
[f] Includes 1 jointly with UK assignee.

What's missing from the patent data presented here?

The data presented here is restricted to stem cell patenting, and also to a relatively short period of time. To properly analyse trends in patenting behaviour and where these are occurring would require a dataset that includes data on patents in areas of regenerative medicine other than

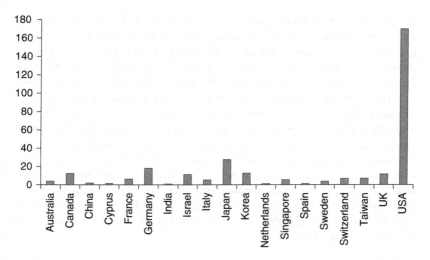

Figure 2.8 Number of patents by country (November 2008–June 2010) [Alternative chart]

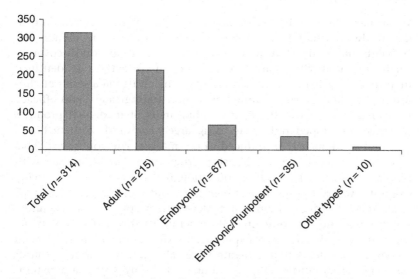

Figure 2.9 Stem cell patents by cell type (November 2008–June 2010)
Source: UKIPO data compiled for UKNSCN.

stem cells, such as tissue and cell culture, wound healing, and so on. This is because it is reasonable to assume patents on the latter types of research would be recorded more frequently in the past compared to recent years. Data collected over a greater time period – say the past 10–15 years – is also required. Also, the data presented here is restricted to patents granted and it would be interesting to analyse the number of patent applications compared to the number granted and to do this according to the major patent offices (EPO, USPO, and others) plus the Patent Control Treaty (PCT). Details on the location of assignees for stem cell patents are analysed above but similar information on other areas of regenerative medicine would be an important resource. Bergman and Graff (2007b), who provide detailed information on patent applications in earlier periods, report that after the PCT, US, and EPO, the most active countries for stem cell filings were Australia, Canada, Japan, Germany, China, the United Kingdom, and Israel.

Is 'Big Pharma' coming to the table? Regenerative medicine and the pharmaceutical industry

What part, if any, is being played by the pharmaceutical industry in the translation of regenerative medicine to the clinic? The conventional model of drug development concentrates on discovering small molecules with pharmacological activity. If activity is identified in animal models the agent is brought to the clinic via a series of clinical and regulatory stages aimed at demonstrating safety and efficacy in humans. Since the 1960s, many hundreds of medicinal products have been developed in this way, using largely chemical synthetic methods, and subsequently introduced to the market. The commercial goal has been to develop a 'blockbuster' drug, defined as a product with annual sales of at least $1 billion (though there are of course many examples where the revenue stream is several times larger). This type of approach is still common despite development pipelines being under severe stress and the conventional model under threat. The 'low hanging fruit' has already been exploited, making it much more difficult to develop products that generate large sales, plus development costs have increased substantially. A number of highly significant patents have either expired or are due to shortly and these patents underpin a large proportion of these companies' current revenue – the so-called 'patent cliff' (Datamonitor, 2010; Dunoon, and Vollebregt, 2010). This situation provides another strong incentive to examine new avenues for maintaining what have historically been high profit levels and big share-holder returns. Companies are therefore looking for alternative models, with an increasing emphasis on targeted treatments. Although smaller in terms of unit sales, 'niche' products can provide larger net revenue because health care providers are likely to be willing to pay more for products that show high efficacy in targeted populations. Orphan drugs are in effect 'niche' products and the proportion of both approvals and revenue that pharmaceutical companies have received from this source has jumped considerably in recent years (Philippidis, 2011; Melnikova, 2012).[30] Many prospective regenerative medicine products will satisfy the criteria for orphan drug status since they are directed at rare diseases.

The evidence to date shows that pharmaceutical companies' interest takes several forms (McKernan et al. (2010). These include direct investment by establishing a specialist unit within the parent company; or ownership of a dedicated regenerative medicine company; or equity capital investment in such firms. Most companies have a venture capital arm and some are investing funds in regenerative medicine companies. Collaboration with SMEs in areas of mutual interest, rather than forms of direct investment, is another approach being adopted. A good example of the first approach is Pfizer's regenerative medicine unit established at Cambridge, UK, in 2010. Pfizer also recently agreed to collaborate with Athersys to develop and commercialise the latter's MultiStem technology for inflammatory bowel disease and is also funding a Phase II trial for ulcerative colitis. MultiStem is a patented and proprietary cell therapy product consisting of a class of stem cells obtained from the bone marrow of healthy adult donors. As such, it is the type of stem cell therapy that 'Big Pharma' is likely to be most comfortable with. As mentioned earlier, Pfizer also has an agreement with UCL on developing a stem cell-based therapy for age-related macular degeneration (AMD), a leading cause of blindness in older people and for which there is currently no effective therapy, plus other related retinal diseases. The company also has an agreement with ViaCyte, which is developing a hESC-based therapy for diabetes, hoping to supply pancreatic cells that make insulin.

Another example of direct involvement is where a company has an existing division or subsidiary in areas where it is a natural progression in therapeutic and technical terms to explore regenerative medicine possibilities. Typical of this type is J&J's medical devices company, De Puy, and its Advanced Technologies and Regenerative Medicine (ATRM) affiliate. In both of these examples, interest has evolved from long-term engagement in tissue engineering and medical devices development. In the case of ATRM, the company is exploring the use of cells and scaffolds to reconstruct organs for the treatment of chronic disease; the systemic or direct allogeneic cell administration to address serious neurological disorders; and combining autologous cells and scaffolds to address moderate to severe arthritis.

A different approach is to work with leading academic research centres. In late 2008 GlaxoSmithKline (GSK) signed a five-year $25 million agreement with the Harvard Stem Cell Institute (HSCI), which brings together several leading centres including the Dana Faber Cancer Institute, with projects in cardiovascular, obesity, oncology, neurology, muscle, and immunology now under way. Described as an 'equal partnership' with a joint steering group, according to one report the arrangement brings a lot of multidisciplinary expertise under one 'virtual umbrella', giving GSK 'one stop shop' access to in-depth knowledge of stem cell biology, leveraged funding, some unique cellular assays, patient populations, and world-class clinical knowledge (Aldridge, 2010). Meanwhile, GSK brings its compound libraries, high-throughput screening, regulatory knowledge, and drug development expertise to the table.

Other companies appear to be more circumspect, with AstraZeneca, for example, adopting a strategy of 'pursuing projects [in-house] and external collaborations... [which are] aligned to our existing areas of disease interest such as respiratory and inflammation, cardiovascular and metabolic and neuroscience.' But AstraZeneca, which currently lacks new products in its pipeline, also has a New Opportunities Group that 'can pursue potentially fruitful regenerative medicine approaches in areas outside the company's core interests, such as bone, muscle and ophthalmic disease' according to a recent interview (Aldridge, 2010). In essence the company's projects reflect their core strengths (i.e., small and large molecule development) and are directed at the modulation of stem cells rather than cell therapy per se. However, the company is interested in understanding the progress being made in cell-based regenerative approaches to disease and again external partners are viewed as essential to this strategy. Like Pfizer, AstraZeneca have an agreement with the Institute of Ophthalmology at UCL, this one directed at diabetic retinopathy. The company also has a long-standing deal with Cellartis AB, focusing in the application of hESC technologies; and involvement with the UK public–private partnership Stem Cells for Safer Medicines.[31]

Another good example is the medium-sized speciality company, Shire, headquartered in Ireland, which recently announced the purchase of Pervasis Therapeutics for $200 million to boost its regenerative medicine unit, adding the small company's cellular therapy, Vascugel®, currently in midstage development for vascular repair in patients on haemodialysis (Crunkhorn, 2012).[32] The buyout follows Shire's $740 million purchase last year of Advanced BioHealing, which specialises in cell-based therapies. However, at this stage it appears that involvement by the major pharmaceutical companies, whilst potentially highly significant, is likely to focus on induced pluripotent (rather than embryonic) cells, and toxicity testing.

While direct industry involvement in stem cell therapeutics looks to be some way off, there has been a noticeable change in the industry's assessment of the potential of gene therapy. Gene therapy – the replacement of a patient's defective genes by healthy ones – was hyped in the 1990s as the 'next big thing' in drug development. A string of high-profile setbacks including patient deaths led to a rapid withdrawal of investment in the field. However, the recent past has seen investment returning. As noted earlier, the REMEDiE project identified some 133 companies active in gene therapy development,[33] whilst a recent Bloomberg report found that 'drug developers big and small' are investing heavily in the field. Among the investors are Genzyme (now part of Sanofi) which historically has concentrated on rare diseases, with an investment of some $200 million over the last decade. Novartis has also been active with a $213 million investment in GenVec, to fund a pre-clinical programme in hearing loss and balance treatments. And Pfizer has negotiated a $145 million agreement to co-develop a pre-clinical hepatitis drug with Tacere Therapeutics that involves a gene delivery mechanism. Gene therapy approaches are also being explored for diseases like Parkinson's and Alzheimer's (Waters, 2010).

Conclusion

Some broad trends in translational activity can be discerned from the data presented in this chapter. With regard to location most activity is concentrated in the United States, which is not surprising given its dominant position in the biosciences and the strength of its capital markets. In Europe, considerably smaller but still important centres are found in Germany, the United Kingdom, and France, and to a lesser extent, Spain and Switzerland. Belgium is also an important player with

Tigenix (now including Spain's Cellerix), arguably the leading European developer of non-stem cell products at present. In Asia, China, Japan, and South Korea continue to devote considerable resources to the field. Data collection issues undoubtedly underestimate the scale of activity in China in particular and probably other countries in this region also, and anecdotal evidence supports this view.

As one would expect given the emerging nature of the field, the bulk of activity is conducted by SMEs across all regions, with a small but growing interest shown by 'Big Pharma'. In the United States, 'hot spots' of activity are concentrated in California, Maryland, Massachusetts, and Florida, plus clusters in Pennsylvania, New York/New Jersey, and Texas. Most of these centres also reflect, and benefit from, efforts by US federal and state agencies to encourage development in the field.

The data on developments in four 'product types', autologous, allogeneic, 'other' products (mainly gene therapy), and services, show that slightly more autologous treatments are under development compared to allogeneic products. However, these two categories are outnumbered by companies engaged in gene therapy approaches. Whilst there is some dispute about the number of actual companies it is clear that substantial investment is again being made in gene therapy approaches, particularly when contrasted with the huge decline witnessed in the late 1990s. However, apart from the recent approval of a product by the Russian regulatory authorities, no gene therapy products are marketed outside China, although many CTs have been conducted or are under way. More than a hundred companies are developing services such as diagnostics for drug development based on iPS technology and cell manufacturing capacity. The service category is likely to be increasingly important within the overall picture in terms of value and commercialisation timeframe.

Efforts to commercialise stem cell treatments concentrate on three therapeutic areas: cardiovascular, gastrointestinal and the CNS. Within cardiovascular, two areas of disease are focused on: CLI and AMI.

There has been a marked increase recently in the number of CTs being conducted. The dominance of the United States is reflected in the fact that most corporate CTs are sponsored by US companies. A small number of non-stem cell products have received regulatory approval and are on the market. As discussed above, there are adult stem cell treatments, mostly autologous but also some allogeneic, undergoing Phase II studies, and Phase III in a few cases. It is therefore possible that one or two such products will enter clinical practice within a few years. There are also potentially highly significant early-stage clinical trials under way using cells from embryonic sources: the first sponsored by Geron for spinal cord injury (although the company has stopped further development) and two Phase I/II trials for Stargardt's disease and other retinal conditions, sponsored by ACT and UCL/London Project to Cure Blindness. Finally there is growing interest in iPS technologies for drug discovery and disease research and, as already noted, these developments are the main interest for pharma companies at present.

More than three times as many patents granted apply to adult stem cells compared to the number referring to hESCs. The number of stem cell patents granted to assignees in the United States far outstrips the number granted to assignees located in other countries, reflecting the current dominance of the United States in both stem cell research and commercialisation opportunities. Universities and other 'non-profit' centres hold a strong position in stem cells patents relative to corporate interests, with the former being assigned more than 50% of granted patents.

Overall, then, this chapter has provided a detailed analysis of the current research and commercial boundaries and trends that characterise the field of regenerative medicine, as measured by company activity, patents, and clinical trials. The analysis points to a complex bioeconomy which is developing on different fronts rather than in some linear, or convergent way. This issue of possible innovation paths is explored more fully in the following chapter.

Notes

1. The chapter does not consider the phenomenon known as stem cell tourism in which severely ill patients travel to clinics around the world where unauthorised stem cell-based treatments are offered, often in the absence of rigorous scientific and ethical requirements. Committee for Advanced Therapies and CAT Scientific Secretariat (2010), Ryan et al. (2010); Dolgin (2010).
2. ATMPs are 'innovative, regenerative therapies which combine aspects of medicine, cell biology, science and engineering for the purpose of regenerating, repairing or replacing damaged tissues or cells', and can be a gene therapy, a somatic cell therapy, or tissue engineered product 'that contains or consists of cells or tissues that have either been subject to "substantial manipulation" or that are not intended to be used for the same essential function(s) in the recipient as in the donor' (Regulation (EC) No 1394/2007). 'The ATMP Regulation' entered into force on 30 December 2007 and applied from 30 December 2008.
3. Such practices are, of course, relatively routine nowadays. As Kemp (2006) notes, once the two technologies of surgical expertise and immunosuppression were united, transplant medicine advanced rapidly, with kidney,

liver, lung, pancreas, and heart transplants successfully undertaken, over the period 1954–1967. The next breakthrough occurred in 1968 with the first bone marrow transplant (see McCann, 2003). Here, the introduction of hemopoetic stem cells, rather than a fully functional organ, 'seeded' the reconstitution of all hemopoetic cells in the patient and a new bone marrow slowly developed.

4. Two easily identifiable 'turning points' are the identification and isolation of hESCs (Thomson et al., 1998), and the creation of iPS by Yamanaka and colleagues (Takahashi et al., 2007). The latter is particularly interesting as iPS cells have the potential of embryonic cells but, being derived from so-called 'adult' cells, avoid the ethical controversies surrounding hESCs.

5. The potential regulatory uncertainties include a lack of suitable animal models, the fact that living cells interact with the environment they are in and have the potential to migrate, and in many cases, a lack of a good understanding of the mechanism of actions, which makes it difficult to satisfy the safety and efficacy standards set by regulatory agencies.

6. Geron is conducting a Phase 1 trial in the United States to evaluate the safety of its hESC-based product candidate, GRNOPC1, a population of living cells containing oligodendrocyte progenitor cells, in patients with thoracic spinal cord injuries. However, the company is not enrolling any additional patients in this study, but patients who have been treated already are being followed for up to 15 years in accordance with pre-established clinical protocols. This was the first study in humans of a hESC product and Geron was one of only two corporations with major investment in hESC development.

7. Dendreon's Provenge (sipuleucel-T), an autologous immunotherapy approved in the United States for prostate cancer treatment, is a good example. Other companies in what is admittedly a definitional 'grey area' include Brucells (Belgium), Molmed (Spain), LKTFarma (France), Boston Biotechnology Inc. (US – now owned by Japan's Dainippon Sumitomo Pharma), and Medinet (Japan).

8. See, for example, Morrison (2012); Brown and Michael (2003); and Brown (2003).

9. California has committed $3 billion over ten years to stem cell research through the CIRM. The UK government-funded Research Councils and Technology Strategy Board are investing £75 million in translational science, part of a strategy for UK regenerative medicine (MRC, 2012). Germany has also made significant investment through its cell-based therapies initiative (€30 million, 2005–2009) and stem cell initiative (€9 million, 2008–2012), and funded a number of translational centres and research 'clusters of excellence' (BIS, 2012).

10. Whitaker and Foley (2011).

11. Note that the category 'Big Pharma' here includes large device, diagnostic, and service companies in addition to pharmaceutical companies.

12. Figures 'prior to 1995' are provided because these firms are still in existence. Under 'Closed', no data prior to 2003 was collected. For earlier data see Martin et al. (2006).

13. Lysaght et al. (2008), for example, contrast the period 2002–2007 with the 'downturn of 2000–2002, at which time tissue engineering was in shambles because of disappointing product launches, failed regulatory trials, *and*

the general investment pullback following the dot-com crash' [emphasis added] (p. 305).

14. The takeover reportedly followed unimpressive results from a Phase III trial of the company's autologous cell therapy, Ontaril, a product based on expanded mesenchymal stem cells obtained from adipose tissue for treatment of complex perianal fistulas in patients with and without Crohn's disease.

15. Other countries have of course adopted this type of approach and built centres which seek to concentrate investment and expertise. The most obvious example is Singapore's Biopolis, created in the early 2000s (Clancey, 2012). There are also examples of this model in Europe, in France and Germany in particular.

16. 'Tissue engineering' (or bioscaffolds) was counted in the 'allogeneic' category.

17. Becton Dickinson have other collaborations, for example, with StemCell Technologies; and the WiCell™ Research Institute, who hold the important University of Wisconsin patent portfolio.

18. Angel Biotechnology is providing the material for the current ReNeuron clinical trial for ischaemic stroke.

19. For example, in Europe a clinical trial must comply with the requirements of the relevant Competent Authority and the EU's ATMP regulation, according to how it is defined.

20. Unfortunately, clinicaltrials.gov was unable to explain why this discrepancy occurs when contacted.

21. The 'hospital exemption' route refers to the fact that some jurisdictions allow exemption from standard controls if the procedure is conducted in a hospital and is small scale.

22. Geron announced that it was concentrating on other products in its pipeline based on small molecule research.

23. The European Court of Justice (ECJ) ruled that stem cell processes which require the prior destruction of human embryos or are based upon the use of human embryos are not patentable (European Court of Justice, 2011a).

24. CLI is the obstruction of the arteries that seriously decreases blood flow to the extremities, resulting in pain, non-healing wounds, and tissue necrosis.

25. AMI, or heart attack, results from interruption of blood supply to the heart, causing cardiac cells, which cannot regenerate, to die.

26. However, in the much larger US market, the FDA is demanding more efficacy data before approval.

27. The assistance of UKIPO and UK National Stem Cell Network (UKNSCN) in providing the source data is gratefully acknowledged.

28. For example, the US PTO requires the criteria to be well-defined with the inventor contributing to the conception not merely reduction to practice (Friedman, 2010).

29. As Friedman (2010) notes, additionally the inventors may not live in the same country as the corporation named as the patent assignee, and development may involve contributions from researchers in multiple countries.

30. Indeed, several products with orphan drug status have sales in excess of $1 billion and hence are 'blockbuster' products by definition. In the United States a disease is considered to be 'rare' and can be considered for orphan

designation if it affects fewer than 200,000 individuals, and in the EU it is defined as having a prevalence of fewer than 5 in 10,000 people.

31. Stem Cells for Safer Medicines is a pre-competitive consortium of UK-based academic groups funded jointly by a number of UK Government departments and pharmaceutical companies, focusing on the development and application of stem cell assays for predictive toxicology and. See www.sc4sm. org for details.

32. Vascugel is an experimental endothelial cell-based drug for boosting blood vessel healing and improving access to vessels in patients with kidney disease who require hemodialysis. Importantly, the product has orphan drug designation from the FDA and EMA.

33. Another report issued in 2010 says there are 189 companies active in the area, with FDA figures showing 254 clinical studies using gene therapy under way in the United States alone (Martino, 2010; Waters, 2010).

References

Aldridge, S. (2010) Regenerative medicine: small steps towards the great leap forward, *InPharm* (16 August). Available at http://www.inpharm.com/news/regenerative-medicine-small-steps-towards-great-leap-forward

Appelbaum, F. R. (2007) Hematopoietic-Cell Transplantation at 50, *New England Journal of Medicine*, 357: 1472–1475.

Bergman, K. and G. Graff (2007a) The global stem cell patent landscape: implications for efficient technology transfer and commercial development, *Nature Biotechnology*, 25(4): 419–424.

Bergman, K. and G. Graff (2007b) *Collaborative IP Management for Stem Cell Research and Development*, Davis, CA: Centre for Intellectual Property Studies (CIP), Göteborg and Public Intellectual Property Resource for Agriculture (PIPRA).

BIS (2011) *Taking Stock of Regenerative Medicine in the United Kingdom*, (July). UK Department of Business, Innovation and Skills (BIS) and UK Department of Health. Available at http://www.bis.gov.uk/assets/biscore/innovation/docs/t/11-1056-taking-stock-of-regenerative-medicine

Blackburn-Starza, A. (2011) Ban on embryonic stem cell patents by European Court of Justice, *BioNews*, 630(24 October). Available at http://www.bionews.org.uk/page_109818.asp

Boseley, S. (2011) Geron abandons stem cell therapy as treatment for paralysis, *The Guardian* (15 November). Available at http://www.guardian.co.uk/science/2011/nov/15/geron-abandons-stem-cell-therapy

Brown, N. (2003) Hope against hype – accountability in biopasts, presents and futures, *Science Studies*, 16(2): 3–21.

Brown, N. and M. Michael (2003) A sociology of expectations: retrospecting prospects and prospecting retrospects, *Technology Analysis and Strategic Management*, 15(1): 3–18.

Clancey, G. (2012) Intelligent island to biopolis: smart minds, sick bodies and millennial turns in Singapore, *Science, Technology and Society*, 17(1): 13–35.

Committee for Advanced Therapies and CAT Scientific Secretariat (2010) Use of unregulated stem-cell based medicinal products, *The Lancet*, 376 (14 August): 514.

Crunkhorn, S. (2012) Deal watch: shire increases focus on regenerative medicine, *Nature Reviews Drug Discovery*, 11: 430.

Cyranoski, D. (2012a) China's stem-cell rules go unheeded, *Nature*, 484: 149–150.

Cyranoski, D. (2012b) Stem-cell therapy takes off in Texas, *Nature*, 483(1 March): 13–14.

Daar, A. S. and H. L. Greenwood (2007) A proposed definition of regenerative medicine, *Journal of Tissue Engineering and Regenerative Medicine*, 1(3): 179–84.

Datamonitor (2010) Pharmaceutical key trends 2010 – the patent cliff dominates but growth opportunities remain, *DataMonitor* 259(17 March).

Deuten, J. J. and A. Rip (2000) Narrative infrastructure in product creation processes, *Organization*, 7(1): 69–83.

Dolgin, E. (2010) Survey details stem cell clinics ahead of regulatory approval, *Nature Medicine*, 16(5): 495.

Dunoon, A. and V. Vollebregt (2010) Can regenerative medicine save Big Pharma business model from the patent cliff?, *Regenerative Medicine*, 5(5): 687–690.

Dutton, G. (2012) Stem cell applications hasten into the clinic, *Genetic Technology and Biotechnology News*, 32(2) (January 15): 1–2.

European Court of Justice (2011a) A process which involves removal of a stem cell from a human embryo at the blastocyst stage, entailing the destruction of that embryo, cannot be patented. Press Release No 112/11, Luxembourg, 18 October 2011. Judgment in Case C-34/10 Oliver Brüstle v Greenpeace e.V. Available at http://curia.europa.eu/jcms/upload/docs/application/pdf/2011-10/cp110112en.pdf

European Court of Justice (2011b) Judgment of the Court (Grand Chamber) of 18 October 2011. Oliver Brüstle v Greenpeace e V. Reference for a preliminary ruling: Bundesgerichtshof – Germany. Directive 98/44/EC – Article 6(2)(c) – Legal protection of biotechnological inventions. Available at http://curia.europa.eu/juris/liste.jsf?language=en&num=C-34/10

Friedman, Y. (2010) Location of pharmaceutical innovation: 2000–2009, *Nature Reviews Drug Discovery*, 9: 835.

Ginty, P. J., E. A. Rayment, P. Hourd, and D. J. Williams (2011) Regenerative medicine, resource and regulation: lessons learned from the remedi project, *Regenerative Medicine*, 6(2): 241–253.

Haseltine, W. A. (2001) The emergence of regenerative medicine: a new field and a new society, *The Journal of Regenerative Medicine*, 2: 17–23.

Hogarth, S. and B. Salter (2010) Regenerative medicine in Europe: global competition and innovation governance, *Regenerative Medicine*, 5(6): 971–985.

Hopkins, M., S. Mahdi, S. M. Thomas and P. Patel (2006) *The Patenting of Human DNA: Global Trends in Public and Private Sector Activity* (The PATGEN Project). Sussex: Science and Technology Policy Research Unit (SPRU).

Kemp, P. (2006) History of regenerative medicine: looking backwards to move forwards, *Regenerative Medicine*, 1(5): 653–669.

Light, D. W. (2009) Global drug discovery: Europe is ahead, *Health Affairs*, 28: 969–977.

Lindvall, O. and I. Hyun (2009) Medical Innovation Versus Stem Cell Tourism, *Science* 324(5935): 1664–1665.

Lysaght, M. J., A. Jaklenec and E. Deweerd (2008) Great expectations: private sector activity in tissue engineering, regenerative medicine, and stem cell therapeutics, *Tissue Engineering: Part A*, 14(2): 305–315.

Martin, P. et al. (2006) Commercial development of stem cell technology: lessons from the past, strategies for the future, *Regenerative Medicine*, 1(6): 801–807.

Martin, P. et al. (2008) Capitalizing hope: the commercial development of umbilical cord blood stem cell banking, *New Genetics and Society*, 27(2): 127–143.

Martino, M. (2010) Gene therapy rises from the dead, *FierceBiotech* (23 April). Available at www.fiercebiotech.com/story/gene-therapy-rises-dead/2010-04-23?utm_medium= nl&utm_ source=internal#ixzz1HRiAK3EG

Mason, C. (2007) Regenerative medicine 2.0, *Regenerative Medicine*, 2(1): 11–18.

Mason, C. and P. Dunnill (2008a) The strong financial case for regenerative medicine and the regen industry, *Regenerative Medicine*, 3(3): 351–363.

Mason, C. and P. Dunnill (2008b) A brief definition of regenerative medicine, *Regenerative Medicine*, 3(1): 1–5.

McCann, S. R. (2003) The history of bone marrow transplantation, in N. S. Hakim and V. E. Papalois (eds) *Organ and Cell Transplantation*. London: Imperial College Press.

McKernan, R. et al. (2010) Pharma's developing interest in stem cells, *Cell Stem Cell*, 6: 517–520.

Melnikova, I. (2012) Rare diseases and orphan drugs, *Nature Reviews Drug Discovery*, 11: 267–268.

Morrison, M. (2012) Promissory futures and possible pasts: the dynamics of contemporary expectations in regenerative medicine, *BioSocieties*, 7: 3–22.

MRC (2012) *A Strategy for UK Regenerative Medicine*. Document prepared by the UK Medical Research Council (MRC) in partnership with the BBSRC, EPSRC, ESRC and the Technology Strategy Board (TSB) (April).

Neish, J. (2007) Stem cells as screening tools in drug discovery, *Current Opinion in Pharmacology*, 7(5): 515–520.

Nelson, J. (1998) Analysis of patenting within gene therapy, *Expert Opinion on Therapeutic Patents*, 8(11): 1495–1505.

Oldham, P. and A.M. Cutter (2006) Mapping global status and trends in patent activity for biological and genetic material, *Genomics, Society and Policy*, 2(2): 62–91.

Philippidis, A. (2011) Orphan drugs, big pharma, *Human Genetics Therapy*, 22: 1037–1040.

Plagnol, A. C., E. Rowley, P. Martin and F. Livesey (2009) Industry perceptions of barriers to commercialization of regenerative medicine products in the UK, *Regenerative Medicine*, 4(4): 549–559.

Reiss, T. and E. Strauss (1998) Gene therapy – the dynamics of patenting worldwide, *Expert Opinion on Therapeutic Patents*, 8(2): 173–179.

Ryan, K. A. et al. (2010) Tracking the rise of stem cell tourism, *Regenerative Medicine*, 5: 27–33.

Suarez-Villa, L. (2000) *Invention and the Rise of Technocapitalism*. Boston, MAUS and Oxford, UK: Rowman & Littlefield.

Takahashi, K. et al. (2007) Induction of pluripotent stem cells from adult human fibroblasts by defined factors, *Cell*, 131(5): 861–872.

Thomson, J. A. et al. (1998) Embryonic stem cell lines derived from human blastocysts, *Science*, 282: 1145–1147.

US Department of Commerce (2010) US Patent and Trademark Office. *Manual of Patent Examining Procedure*. Virginia: US Patent and Trademark Office.

Wainwright, S. et al. (2008) Shifting paradigms? Reflections on regenerative medicine, embryonic stem cells and pharmaceuticals, *Sociology of Health and Illness*, 30(6): 959–974.

Waters, R (2010) Gene Therapy revival spurs hope for Genzyme, Pfizer, *Bloomberg Business Week* (26 April). http://www.bloomberg.com/news/2010-04-22/-dead-as-doornail-gene-therapy-revival-spurs-hope-for-genzyme-pfizer.html

Webster, A. (2013) Introduction: the boundaries and mobilities of regenerative medicine, in A. Webster (ed) *The Global Dynamics of Regenerative Medicine: A Social Science Critique*. Basingstoke: Palgrave Macmillan.

Webster, A. et al. (2011) Regenerative medicine in Europe: emerging needs and challenges in a global context (REMEDiE), Final Report to the European Commission, SATSU, University of York. Available at http://www.york.ac.uk/satsu/remedie.reports

Whitaker, M. and L. Foley (2011) What are the challenges of bringing RM to the clinic? Paper presented at 3rd REMEDiE conference, Bilbao, Spain, 10 April. Available at http://www.york.ac.uk/satsu/remedie

Yamanaka, S. (2009) A fresh look at iPS cells, *Cell*, 137(April 3): 13–17.

3
Biocapital and Innovation Paths: The Exploitation of Regenerative Medicine

Michael Morrison, Stuart Hogarth, and Beth Kewell

Introduction

This chapter examines the dynamics of innovation in regenerative medicine (RM), focusing specifically on the contemporary activities of European-based RM firms. For the purposes of this chapter the term 'innovation' is taken to refer to the development and deployment of novel technologies. As Chapters 1 and 2 have shown, 'regenerative medicine' is a heterogeneous domain incorporating a range of technologies. Different technological options within RM can be complementary or in competition with one another depending on how their deployment is envisaged. The heterogeneity of RM is not limited to specific material technologies, but also encompasses different models of how its products might be delivered in the clinic, how they might be reimbursed under different financial health care regimes, and different aims in terms of the diseases and patient populations particular RM technologies are best suited to address. There are thus different pathways which innovation in RM can potentially follow.

The range of visions of the field can be examined in terms of industry change over time, from somatic cell-based tissue engineering (TE) to stem cell-dominated RM (Mason, 2007); autologous versus allogeneic approaches to cell therapy and their accompanying business models; and tensions between the potential of RM as a (routine) clinical service and the economic/ideological drivers focusing on the industrial product-orientated aspect of RM. In regard to this latter point, it is important to consider that innovation does not take place in a vacuum. Rather, the actions of firms, scientists, investors, regulators, and so on are all locatable within a broader global domain of RM, and within particular political and economic regimes which exert a strong influence on the nature (and form) of technological innovation itself. This chapter will thus begin by placing European RM within the framework of the 'knowledge-based bioeconomy' in order to address the question of whether the policy emphasis on developing cell-based RM/TE products through a fully commercialised bioeconomy model contributes towards European biomedical innovation or works against it. Accordingly, the benefits and problems – especially in view of the post-2008 economic climate of 'austerity' of this innovation model – will be discussed and an alternative, in the form of the 'hidden innovation system' of hospital-based clinical development described by Hicks and Katz (1996), introduced before presenting an overview of the key contemporary visions for RM.

Drawing on a range of empirical data collected as part of the REMEDiE project (including financial analysis of the RM sector, interviews with leading European RM industry figures, and detailed analysis of European RM firms) the commercial performance of the RM industry and the technological choices made by European RM firms are then analysed to test the fit between industrial visions for the field and current patterns of development. The importance of this evaluation is not only to characterise the European RM field, but also to identify key drivers that have shaped its development including economic, governance, historical, and other factors. This in turn allows an informed consideration of the future prospects and innovation trajectories for RM in Europe. Economic considerations are paramount in a bioeconomy model of innovation and the current risk-averse investment climate poses serious risks to the sustainability of a European RM sector of any size. It is in this context that alternative innovation paths, better suited to the complex material, temporal, and socio-political aspects of developing living human tissue-based therapies become a realistic and worthwhile consideration.

Bioscience innovation in context

In order to properly consider the commercial development of RM in Europe (and beyond) it is necessary to locate the activities of both individual firms and national RM industries in the wider context of the bioeconomy. Broadly, the bioeconomy is a particular life sciences-orientated formulation of 'knowledge-based economy', in which innovation is positioned as the source of raw value and the new knowledge it produces, when codified in the form of intellectual property rights (IPR), becomes the foundation for new rounds of economic activity and trade (Cooke, 2001). In this model, small 'start-up' firms and companies 'spun out' from universities have been considered the best vehicles for the commercialisation of innovative technologies. Concomitantly, technoscientific research and development has become increasingly central to national and international economic operations and has been subject to increasingly strategic management and future-orientated regimes of planning (Borup et al., 2006; Kewell and Webster, 2009). Leydesdorff and Etzkowitz (1996) theorised that innovation in a knowledge economy could be characterised as a 'triple helix' of links between the state, the academic sector and industry, where the state supports innovation through market-promoting reforms and (strategic) funding of science and technology R&D in the academic sector. Promising research from the academic sector is then patented and spun out into small biotech firms or licensed to entrepreneurs setting up their own companies. The subsequent development of research knowledge into products and services is then largely dependent on companies' abilities to secure financial support from venture capital (VC) firms and other investors.

This economic linkage between science and speculative investment – the 'mangle of science and capital in the 21st century' (Tutton, 2011, p. 412, after Pickering, 1995) – has been described as the 'privatisation of science' (Mirowski, 2011), technocapitalism (Suarez-Villa, 2009), and 'venture science' (Dumit, 2003). While science and technology have always been, to an extent, future-orientated enterprises, promising particular benefits to human society, it can be argued that the increasingly strategic economic management and highly future-orientated nature of speculative investment make contemporary science and technology development qualitatively different from what has gone before (Borup et al., 2006). Knowledge economies can be characterised by the notion of a 'double promise' – where the value of intangible knowledge in the present is closely intertwined with both the projected social benefits arising from new technologies and the associated promise of future economic growth and increasing returns on speculative capital (Morrison and Cornips, 2012). Hope (2009) has described this economic situation as 'a collapse of the future into the present', (p. 68) as the value (share price) of investment-dependent companies, such as biotech firms, ceases to be a reflection of past profitability and instead becomes an estimate of a firm's projected future profits as estimated by market analysts and other expert voices.

The bioeconomy extends this model into the realm of the life sciences and specifically biotechnological control and manipulation of (often living) organic biomaterial. Waldby and Mitchell (2006) describe the emergence of global 'tissue economies' of which the RM industry is partly based on the procurement and commodification of human tissues and body parts as the basis for novel medicinal and biotechnological products. Bioeconomic processes involve more than just collection and exchange of biomaterial; the technological manipulation of cells and tissues which is considered to transform them from 'products of nature' into patentable, fungible 'bio-objects' constitutes both reconfiguration and reproduction of the biological (Cooper, 2008). Control of the regenerative capacities of living cells yields 'biovalue', a corporeal surplus which can become the object of speculative investment or 'biocapital' (Kent et al., 2006; Rajan, 2006; Waldby and Mitchell, 2006). In essence, the generative potential of biological life becomes intertwined with that of capital.

The very novelty of technologies based on human cells and other novel biomaterials makes them a form of disruptive innovation. Unlike steady or stable technological innovation, in which incremental improvements are made to existing technologies already established in networks of production, regulation, demand and supply, RM technologies are novel and do not align with existing practices, infrastructures, regulations, and so on (Geels, 2002). While they are often considered to hold the greatest potential for radical change, disruptive technologies initially face a struggle to develop precisely *because* they are 'misaligned' with existing infrastructures, practices and markets, and so on. In order to survive and become established they require the generation of new networks and recruitment of novel or existing partners including investors and government actors, to support the product development and deployment processes. One way in which this can be achieved is through the communication of technological expectations (Borup et al. 2006; Brown et al., 2000).

Expectations are hopeful future-orientated claims about what a nascent technology might achieve if its development is adequately supported. The importance of such visions about novel technologies is that they are not 'just hype', but are actually *generative* of new technoscientific projects. Visions and expectations act to structure and guide actions (technical and otherwise), generate interest, and mobilise resources in support of novel technology development projects, including, crucially, attracting speculative investment from VC firms, angel investors, and even public–private partnerships (Van Lente, 1993; Brown et al., 2000;

Birch, 2006; Borup et al., 2006). The range of extant visions for RM is thus a useful site to assess the different innovation pathways currently being presented, and followed, by European RM firms and will be outlined in subsequent sections.

The hidden innovation system

The triple helix/bioeconomy model, although dominant in biotech policy circles, is not the only model of an innovation network. In a study of the UK science base carried out in the mid-1990s, Hicks and Katz (1996) observed a 'hidden' system of biomedical innovation based in hospitals and public laboratories and funded by research councils and charities, existing alongside the more visible university–state–industry network. Within the 'triple helix' model the health care system is often presented as a barrier to innovation, with sceptical clinicians and cost-conscious health care managers unwilling to embrace new technologies. However, the hidden innovation system presents an alternative account, which sees the health care system as an agent of, rather than an obstacle to, technological change. Focusing on the rapidly growing molecular diagnostics sector, Michael Hopkins (2006) has described the role of this hidden innovation system in developing clinical cytogenetic testing in the United Kingdom. Firms, while not excluded from this innovation system, are significantly more peripheral actors in these networks of technology development, relying instead on 'public sector clinicians and scientists to establish and stabilize this market first' (Hopkins 2006, p. 270).

There is also precedent for 'hidden innovation' in the field of RM. The application of bone marrow transplants, as a therapy for acute blood cancers (especially leukaemia), draws its regenerative effect from the transfer of haematopoietic (blood-forming) stem cells (HSCs) contained within the marrow. Bone marrow/HSC transplants are currently considered the only stem cell therapy currently in routine clinical use (Brown et al., 2006; Daley and Scadden, 2008). In the 1980s and early 1990s attempts to commercialise HSC technology were made by a now largely forgotten 'first wave' of stem cell companies, including Aastrom Biosciences, Amcell, Applied Immune Sciences, CellPro, Progenitor, and SysTemix in the United States and CellGenix in Europe (Martin et al., 2006). However of these seven companies, only Aastrom Biosciences is still active in the stem cell therapy field and none of the firms evolved a truly successful business model for generating a viable long-term revenue source from HSCs as a cancer therapy product (ibid.). Instead, the successful deployment of HSC therapy has largely occurred in a hospital system-based innovation environment. However, it is important to note that bone marrow transplants were not an immediate success. Brown et al. (2006) report that

> [t]he therapeutic promise of bone marrow infusions did not translate into early clinical success; instead, early optimism was confounded by recalcitrant 'material resistances'. By the mid 1960s, BMT was viewed by many as a clinical failure. (p. 336)

Instead of the expected linear progression from 'bench to beside' the routine application of HSCs involved a number of unsuccessful attempts at clinical application followed by subsequent periods of further laboratory investigation. For bone marrow transplant (BMT) treatment to be rendered viable required recognition of the role of the immune system in mediating the transplant of biomaterials between bodies. For HSC transplant to become routine required further cycles through the clinic and the laboratory, including development of technological means to detect and isolate HSCs from the heterogeneous bone marrow tissue (ibid.). This suggests that, at least in some cases, innovation processes requiring long-term (over 30 years in the case of BMT) highly complex, uncertain, and non-linear development pathways are better suited to the clinic rather than the biotech firm.

While the hidden innovation system has its own limitations – Hopkins suggests that in the early part of its innovation journey cytogenetics could be considered underregulated – it is important to the consideration of RM in this chapter as an alternative to some of the difficulties facing the firm-led commercialisation of RM products. This is particularly pertinent in view of the current economic obstacles to the VC-driven bioeconomy approach.

RM and the market

The year 2008 was arguably an *annus horribilis* for the world economy that has inflicted lasting structural damage upon the biomedical and gene-related innovation sectors of Europe and the United States (Ernst and Young, 2009; Gruber, 2009; Guertin, 2009; Martin et al., 2009). The economic 'crash' experiences in the third quarter of 2008 can be economically characterised as a transition from a 'Bull market' to a 'Bear market'. Contextually speaking, the worlds of the Bull and the Bear could not be more contrasting: one represents financial plenty, the other fiscal famine. Significantly, any areas of company performance inefficacy that may have been cushioned in times of plenty and relative operational normativity become threats of magnitude to the survival of firms (Linsley and Linsley, 2009; Taylor, n.d.). Thus, the period prior to 2008 can be regarded, in retrospect, as one of relative calm before a climatic storm, when the problems faced by RM firms were those concomitant with a young innovation industry experiencing growth pangs at the first and second stages of a well-documented biotechnology innovation lifecycle (Cooke, 2001, 2004; Owen-Smith et al., 2002; Coenen et al., 2004; Tait, 2007; Gruber, 2009; OECD, 2009).

Most significantly, the financial position of RM firms in 2008 seems predicated on inward investment and 'new cash' entering the sector rather than a story of productivity and profitability emerging amongst leading enterprises. Bear Markets tend to lead to diminished VC availability and cash poverty. Cash comfort zones disappear and enterprises are forces to rely instead for their capital intake on gearing arrangements with private banks. Firms without access to gearing are likely to represent unsustainable businesses going forwards, because there is a strong possibility currently that the public investor market may mark a phased retreat from the RM sector. These will be dependent to a great degree upon the outcomes of current clinical trials in the United States and South East Asia. Sponsorship solutions may need to be arranged therefore (i.e., between national and/or federal governments and private banks) if small RM firms and new start-ups are to be encouraged under recessionary conditions.

Financial analysis of the RM industry strongly suggests that the RM sector was experiencing important performance vulnerabilities *prior* to the third quarter Bear Market crash of 2008; and that, beyond a relatively small constituency of semi-profitable firms, the high financial risk burdens carried by many publicly listed companies were a threat during the good times but have become more so since. Indeed, it is possible that the presence of severe levels of indebtedness amongst a core of RM firms threatens the lifecycle sustainability of the share-owner-led side to the industry. Despite the existence of 'Angel Shareholders' willing to make high-risk investments in novel, but uncertainty-ridden enterprises under the conditions of relative Bull Market prosperity, it is unlikely that RM enterprises will be able to attract Bear Market investors – they will simply appear too high a risk as an investment vehicle. RM may therefore have to rely on private non-stock-market routes in the near future to find equity and the public and charitable purse for donations and funding. Inevitably, the shrinkage seen in 2001–2002 (following the so-called the 'dot-com' investment crash) appears to be being repeated in the United States and worldwide. Indeed, evidence for this has already being reported in industry digests (Marks and Clerk, 2009; Marrus, 2009). Policymakers and other state actors must now consider how they target support for particular products, technologies, and therapeutics under development within the industry at the present time based not only on the clinical utility of the innovation in question but also in relation to the financial viability of the company orchestrating its production. Firm indebtedness represents a significant barrier to continued innovation in the current climate. It is also possible that future investors, whether private or public, will change their investment behaviours (Barber and Odean, 2001; Hirshleifer, 2001; Redhead, 2008) considerably in the near future: targeting monies at low-risk, high-certainty areas of the science base (i.e., stem cell-based toxicology testing and pharmacogenetics) which are more likely to generate a return on income and with this a rapid social or clinical utility than those involving revolutionary patents and difficult to trial techniques. This may, in itself, curtail the onward development of radical forms of RM under the more conservative, inauspicious, and cash-poor conditions of the Bear Market.

In our exploration of these issues with those working in the corporate sector, the prevailing financial climate especially the lack of private finance has been of concern:

> Well it's not becoming a problem, it's been a problem for several years and it's been a problem for the last decade I would say [...] I think if you were a stem cell company in early 2000 you probably could have you know got some, raised some cash on the market but from you know 2004 onwards really it started to become much more difficult for small biotech companies working on stem cells to get cash.
>
> (UK interview, RM executive)

Another pharmaceutical firm executive agreed with the view that VC finance for RM companies began to dry up in 2004 and several interviewees suggested that there was now no start-up capital available for new RM companies in the European Union (EU). The lack of finance was attributed to investor caution, with RM being viewed as 'too difficult and too far off...full of regulation [and] ethically difficult' (pharma executive). Furthermore, more well-established firms which had received funding feel that they are at a significant competitive disadvantage to their US rivals. One leading EU firm suggested that their main US competitor was 'playing in a different league' with five to ten times the amount of VC funding. One VC executive emphasised the disparity in growth funding, suggesting that European VC firms can build a firm to a €50M valuation but are then forced to sell because they lack the resources to continue to the next stage.

This lack of capital could have a number of consequences for commercial strategy; one interviewee described companies being forced into premature decisions about their pipeline for commercial rather than scientific reasons:

> It then meant that in terms of the cycle evolution for these small companies they were probably forced to make you know key decisions too early in the cycle... like go[ing] to a public listing to get cash. Maybe entered into clinical trials too soon and then you know once you are... in the public gaze you know you are so much at the behest of the investors.

This point speaks to one of the problematic aspects of the knowledge economy model of innovation when applied to the life sciences. Biotech products have notoriously long lead times (often estimated at ten or more years from project inception to market launch) meaning there is a considerable time span throughout which firms must seek investment based largely on the potential of their nascent technologies. This timeframe is somewhat at odds with that of VC firms who are the major investors in RM and other biotech companies, at least during Bull Market conditions. In general VC firms raise money from private institutional and individual investors which is then organised into specific funds, each one with a planned lifetime of between five and ten years (Bottazzi and Da Rin, 2002). Importantly, VC finance does not commit to meeting a company's full financial requirements upfront but provides limited term capital either for a particular round of financing or through a milestone payment contract (Cuny and Talmor, 2005). In round financing a company solicits funding and one or more VC firms can respond by providing a negotiated amount of money intended to sustain a company through a particular developmental stage (e.g., seed or start-up finance). There is no commitment to any further financial support outside the period of the investment and so companies must go through several rounds of seeking investment support throughout their lifetime. In contrast, milestone payments offer a longer-term, predetermined level of financial commitment, but one which is contingent on a company's performance, in this case adjudged by achieving certain specified targets such as appointing key staff members of suitable experience, developing a product prototype, or generating promising data from a clinical trial.

The periodicities of VC finance compel a relatively short-term future orientation for RM firms, as capital raised through VC funding must be employed to meet or beat short-term expectations for performance in time for the next round of funding, or to achieve the next milestone, in order to secure the capital needed to sustain the company for another period until the company achieves some measure of self-sufficiency or other 'exit'. Ultimately VC firms have a 'liquidity preference'; that is, they hope that a proportion of the companies backed by a given fund will achieve success by being floated on the public stock market – known as an initial public offering (IPO) or by being bought over by a larger competitor (Bottazzi and Da Rin, 2002). Thus successful RM firms must not only generate hopeful expectations about the potential of their technologies in order to generate initial support from investors and other actors, they must continuously produce positive news about their progress from basic research to product launch as long as they are reliant on securing investment to remain active (Morrison and Cornips, 2012). It is for this reason that future-orientated promissory narratives are paramount in the bioeconomy and why the different visions for commercial RM technologies are such a useful site for studying the dynamics of innovation in the RM sector.

Contemporary visions of RM

By its own (multiple) chronologies RM has existed as a concept (and a label), since at least the early 1990s, when the term 'regenerative medicine' itself is widely reported to have been coined (Lysaght et al., 2008). Although it remains a disruptive technology, a certain consolidation of approaches can be detected with core elements becoming established (Van Merkerk and Robinson, 2006). The therapeutic application of living human cells is such a core element, with human stem cells being widely considered the most important, and most promising, category of cells. At the same time, there remains a degree of fluidity about whether potentially compatible technologies, including biodegradable 'scaffolds' to guide and stimulate cell regrowth, genetic modification of cells, and cloning, properly belong under the umbrella of RM and if so, in what capacity. Human cells, the primary biomaterial involved in visions of RM are also rhetorically separated by source, degree of modification, and ability to differentiate, with different technical, therapeutic, and economic expectations associated with different 'types' of cells. This section will present and discuss some of the major dynamics in contemporary visions of RM, hierarchies of cell potential, autologous versus allogeneic cell therapies, and TE versus RM. Finally, the particular difficulties of establishing safety and efficacy through clinical trials associated with RM products will be reviewed.

Hierarchies of cell potential

Many accounts of RM propose a hierarchy of human cell types, often ordered in terms of cell potency – that is, the capacity to which particular cell types are capable of transforming (or being transformed) into other cell lineages. Human embryonic stem cells (hESCs), classified as 'pluripotent' or capable of differentiating into almost every possible cell type found in the human body, are often considered the most promising:

> Non-ESCs are lower in the stem cell hierarchy, [t]hey are thought to have lost the pluripotent capability that ESCs have.
>
> (Bajada et al., 2008, p. 172)

The promise of hESCs lies not only their malleability, but in their apparent capacity to continue to replicate indefinitely in culture, generating a source of potentially inexhaustible biological and economic value. 'Non-ESCs' include 'multipotent' foetal and umbilical cord blood-derived stem cells and adult stem cells which are generally only considered able to differentiate into the cell types of the specific tissue or organ in which they are normally found. The announcement in 2007 of a viable technique to 'reprogram' mature somatic cells to resemble immature, pluripotent cells with properties similar, though not identical, to embryonic stem cells has challenged the traditional hierarchy of cell potency (Takahashi et al., 2007; Yu et al., 2007). For some commentators, induced pluripotent stem (iPS) cells have thus come to replace hESCs at the pinnacle of the hierarchies of cell potential.

This illustrates one set of dynamics at work in contemporary visions of RM. It is considered desirable in some quarters to avoid the use of hESCs because of the ethical problems associated with their derivation. Medical involvement with embryos has historically been a source of considerable public and governmental concern through debates on assisted reproduction and abortion, and the technological development of hESCs, whose derivation currently requires the destruction of a human embryo, has often become an extension of these already highly polarised discourses (Bahadur et al., 2010). At the same time, iPS cells are not an unproblematic option for replacing hESCs. Initially the reprogramming of cells to become pluripotent was done using modified viruses raising concerns that such a technique could never be sufficiently safe for human use. Novel methods of inducing pluripotency in mature cells without employing viruses have since been developed, but the technology is still at a very early stage and there are significant questions about what iPS cells 'actually are', what properties they genuinely display in the long term, and whether they can be controlled on a sufficient scale to make them a viable therapeutic and commercial prospect (Belmonte et al., 2009; Hu et al., 2010). In this context, adult, bone marrow-derived stem cell lineages which have been in clinical use for many years in the form of bone marrow transplants are considered among the best characterised stem cell lineages and, with much less uncertainty about how these cells behave when implanted into a human patient, are also perceived as the safest stem cell type for near-term clinical application (Daley and Scadden, 2008; Martin et al., 2008). Umbilical cord blood-derived stem cells, being widely banked and arguably less painful to donate, have also emerged as a viable clinical alternative to bone marrow and a socially acceptable source of stem cells (Brown et al., 2011).

Tissue engineering and regenerative medicine

Somatic, mature, or 'adult' (non-stem) cells have fixed types and properties and generally cannot differentiate into any other cell type. Approaches drawing on somatic cells to repair the body are often described as TE. Some contemporary commentators on the RM industry have posited that TE has ultimately not been a commercially successful venture and has effectively been superseded by an RM industry based on the potential of pluripotent stem cells. Chris Mason's 'regenerative medicine 2.0' (2007) is an exemplar of this vision of RM (see also Kemp, 2006; Lysaght et al., 2008; Nerem, 2010). While these accounts of RM are dominated by stem cell technologies, a number of alternatives remain. It is possible to detect a set of contemporary visions of RM that eschew explicitly hierarchical definitions of RM in favour of a goal-orientated characterisation of the field, arguing that

> [t]he main defining feature of regenerative medicine is not the use of a specific technology, but rather the goal that brings diverse technologies together: to restore impaired anatomy and physiological and biomechanical function.
>
> (Daar and Greenwood, 2007, p. 181)

These accounts present a more holistic conception of the field where TE approaches, including biodegradable scaffolds to support cell regrowth, sit alongside various types of stem cells, combined cell-and-gene therapies and other options (see Atala, 2007; Gardner, 2007; Sheyn et al., 2010). Even here there is gradation, as different technologies within RM are considered appropriate to different therapeutic challenges. TE approaches are generally aimed at applications such as cartilage and skeletal repair and surface wounds and lesions which have comparatively small patient populations, while hESCs and iPS cells are positioned as offering potential cures for several highly prevalent, chronic conditions such as diabetes and heart disease. This obviously involves very different calculations about the long-term economic value of investing in different RM technologies, especially for national governments interested in addressing the cost to health care systems of major chronic illness.

Autologous versus Allogeneic cells

Autologous cell therapies use cells derived from a patient's own body as the basis of the treatment. This often involves the extraction of cells and an ex vivo step of expanding the population of cells, using a particular medical device or protocol, before re-implanting them at the treatment site. In a number of ways this process resembles an individualised surgical procedure much more than a drug therapy aimed at a particular patient population. As the cells used are the patient's own and the application is a medical procedure there is little scope for IPR in the cell removal and re-implantation process itself. This means companies developing autologous cell therapies must seek to realise financial returns through proprietary rights on the ex vivo parts of the procedure such as the cell culturing and reagents, any devices which are employed, either cell scaffolds or means for expanding and/or differentiating the cell population. Autologous cell therapies also have limited potential to scale up the treatment process since each patient will need their cells expanded ex vivo in isolation (i.e., at a discrete workstation) to avoid cross-contamination risks (MRC, 2012).

Allogeneic cell therapies aim to develop an 'off the shelf' cell therapy product which uses cells derived from a single donor. These cells are treated and grown in culture to give a biomaterial that can be shipped to multiple sites and implanted in patients unrelated and unconnected to the original donor. This model conceives of cell therapy products as much closer to conventional drugs in that one product is suitable for a comparatively sizable population. It has been argued that allogeneic cell therapies offer a much greater scope for profitability in that the cell product can itself be patented and sold on a 'fee per unit' basis like a conventional pharmaceutical product (Whitaker, 2011). This 'cells as drugs' approach is felt to be viewed favourably by investors and other stakeholders as a long-term goal of RM. However, development, accreditation, and standardisation of large-scale automated cell culture for clinical-grade applications remains a work in progress and continues to present significant technical and regulatory challenges for allogeneic products in the short to medium term. Allogeneic cell therapies are often perceived as riskier than autologous ones because the former, as modified 'foreign' biological material, presents a greater risk of immune rejection by the host (patient) and requires greater use of immune system suppressing drugs to work. However, the nature of this divide is not universally accepted, as immunological work has suggested that those cells and tissues thought to be immune privileged, including autologous cells and hESCs, can still present transplant rejection problems (Fairchild et al., 2004, 2007).

Cell therapies and clinical trials

Randomised control trials (RCT) are considered the 'gold standard' method for evaluating the safety and efficacy of pharmaceuticals for human use due to their statistical power to infer the generalisability of results to large (patient) populations (Webster et al., 2011). Pharmaceutical clinical trials are organised in a series of test administrations of the drug being evaluated to increasing groups of patients/volunteers. Phase I tests involve administration of the drug to a limited test group to determine safety and appropriate dosage for subsequent phases, while Phase III trials involve administration to much larger groups of human subjects, ideally using double-blinded administration of a placebo as a control mechanism, and form the basis of data that will be used to determine approval of the intervention by regulatory authorities. However, the use of living human cells as therapies presents a number of challenges to the conduct and organisation of the clinical trial model as currently utilised for pharmaceutical product evaluation.

Many of these difficulties involve the material differences between living cells and 'regular' inorganic or biological drugs. Living cells can grow, replicate, and, in the case of stem cells, differentiate into other cell types. Cell growth and differentiation are complex processes and are difficult to control both in culture and within the body.

The difficulty of tracking or controlling cell behaviour in vivo raises notable safety concerns. While it is true that pharmaceuticals or vaccines cannot be 'recalled' once administered, a timescale for elimination of the drugs or other materials from the body can usually be calculated using standard pharmacological tools; with cell therapies there is considerable uncertainty about how long individual cells will persist within the body and whether they will stay localised to a particular site or move around to engraft at multiple locations. A recent policy document developed by the UK Technology Strategy Board and selected Research Councils notes that

> the very regenerative potential that makes stem cell treatments appear so promising is also the quality that makes them risky: securing stable implantation can be difficult, cell batches can vary over the course of a trial and endpoints might be difficult to determine where patients carry a range of co-morbidities.
>
> (MRC, 2012)

With stem cells there is also the additional concern that their ability to self-replicate and differentiate could lead to the formation of tumourous growths as the ability to proliferate freely is also a characteristic of cancer cells (ISSCR, 2008).

Another significant difference is that cells, unlike industrially produced pharmaceuticals, must be derived from the patient's own body or from a donor (the donor may be another person, a 'spare' embryo or a cadaveric foetus). Donors, and the cells derived from them, exhibit considerable variability, including at the molecular level, and the work on standardising procedures of donation, material collection, and characterisation require considerable work, which remains ongoing (ISSCR, 2008). This adds another layer of variability to the behaviour of cells even before they reach the clinic. The heterogeneity of cells in culture is likely to manifest as heterogeneity of characteristics between different batches of the same cell treatment, which has implications for evaluating the efficacy and behaviour of a given cell therapy in human subjects. It may not be clear which observed effects result from variation in the patient (e.g., different co-morbidities) and which are due to variability in the cell batch. This is particularly a problem for autologous cell therapy trials where, in effect, intervention operates on a 'one batch, one patient' basis. Allogeneic cell therapies with their standardised 'cells as drugs' operation fit much more closely to the pharmaceutical model of clinical trials (Webster et al., 2011). However, because of the additional modification required to standardise or 'purify' cells for allogeneic use there is a considerable early workload in documenting the behaviour of these cells to ensure regulatory authorities that they meet the required safety standards.

Innovation pathways of European RM firms

The European RM sector comprises some 112 companies, with cell therapy firms being the largest single segment. There are also significant numbers of firms operating in the areas of bioscaffold production and service provision. The majority of firms in this latter group, RM service companies, target the academic and commercial stem cell research domains, including provision of human tissue (stem cell lines and some somatic cell types used in TE or drug toxicity screening) and on RM-specific tools and reagents (human stem cell-specific culture media, specialised bioreactors for three-dimensional TE culture, etc.). This predominance of cell therapy-orientated activity matches the discursive emphasis on the centrality of human cells articulated in most contemporary visions of RM. A total of 51 cell therapy firms are located in Europe (2012), with 65 cell therapies available or in development. For the purposes of this analysis, the category of cell therapy firms includes companies working on genetically modified or otherwise treated cells, as they still primarily involve the application of human cells for therapeutic benefit. In terms of the particular innovation pathways favoured by European cell therapy firms, two significant dynamics are readily identifiable. Firstly, the surveyed firms are split almost evenly between those using somatic cells and those developing therapies based on stem cells. Secondly, there is a strong emphasis on the development of autologous rather than allogeneic therapies.

Overview of the industry

The available cell therapies are overwhelmingly based on autologous somatic cell therapies while the cell therapy pipeline shows much greater investment in stem cell technologies and in the development of allogeneic approaches to cell delivery. This, along with the presence of more than 20 dedicated bioscaffold firms, strongly suggests that European RM industry is stratified in a way that reflects the wider history of the RM industry. The available cell therapies and bioscaffolds are mainly TE era procedures using autologous epithelial or cartilage cells to repair skin lesions and restore cartilage damage in joints. Investment in stem cell therapies is often more recent and has opened up possibilities for allogeneic as well autologous approaches, although the former is still heavily favoured by European firms. It is largely the stem cell therapies that are being developed for the more ambitious clinical indications; those that reflect the promise of long-term health care cost reductions through RM, such as cardiac repair, neuroregenerative treatments, and treatment of autoimmune diseases. This suggests that, rather than a wholesale replacement of TE-type approaches by stem cell therapies envisaged in the 'regenerative medicine 2.0' narrative, existing and novel technological components of the RM field coexist in Europe. Kent et al. (2006) suggest that there are also broader cultural and socio-political factors favouring the predominance of autologous cell therapies within the European context:

> An emphasis on autologous products may be seen as commensurate with certain aspects of the political and ethical culture of the UK and Europe, a culture where there is antipathy towards the commercialization and commodification of the body. (p. 14)

In this sense, the understanding of autologous cell therapies as involving 'self-repair' are seen as less transgressive – to individual bodies, identities, and other boundaries – than allogeneic 'off the shelf' approaches and are therefore less politically problematic to produce and regulate. Although the TE-type products generally have smaller patient populations and are considered lower down the hierarchy of potential, in many cases it is the performance of these older, 'less risky' products against which pluripotent stem cell therapies must demonstrate their greater efficiency and cost–benefit outcomes if the latter are to become embedded in routine medical practice in the long term.

Governance and regulation

In addition to temporal and cultural factors, the heterogeneity of approaches and stratification in terms of 'TE versus RM' dynamics can partly be explained in terms of the differential barriers to commercialisation facing different technological options. Bioscaffolds, for example, have the least demanding regulatory pathway of the RM technologies covered in this chapter, as they are classed as medical devices. Bioscaffolds are also easier to produce and are among the oldest commercially available RM products. By contrast, the novelty of human cell-based therapeutic products has historically been the source of regulatory uncertainty and a major source of dissatisfaction among those trying to commercialise such products (Faulkner et al., 2003; Faulkner, 2009; Plagnol et al. 2009). As a response to this uncertainty and the fragmented and conflicting regulatory responses of individual member states, the European Medicines Agency (EMA) developed a centralised framework for the regulation of TE and RM products within Europe. Regulation (EC) No 1394/2007 created the EMA Committee for Advanced

Therapies (CAT) as a specific subgroup of the central regulatory agency to deal with RM and other novel or hybrid technologies and introduced the category of 'advanced therapy medical products' (ATMP) to cover products intended for human use based on gene therapy, somatic cell therapy, or TE.

The ATMP regulations came into force in 2008 and, while no stem cell therapy has yet received a marketing authorisation approval (MAA), the EMA reports that some 48 therapeutic technologies have been classified as advanced therapy medicinal products by the CAT including at least seven stem cell-based therapies.[1] The most notable product to receive approval under the ATMP pathway to date is *ChondroCelect*, an autologous non-stem cell product for repair of damaged cartilage developed by TiGenix NV of Belgium, which received an MAA in June 2009. This is not necessarily a pessimistic or surprising state of affairs as much of the European commercial research on stem cell-based products is currently in the early stages of clinical development. Indeed, the centralised EU approval procedure under the ATMP regulation and the European Medicine's Agency approach to licensing appear to be welcomed by those in the corporate sector. One RM executive suggested that the approval of the first ATMP produced demonstrated that EMA was being 'very lenient'. However, there are also concerns that the regulatory framework is still evolving and that not all aspects of regulation have been harmonised. The regulation of clinical trials by different member states is one such area and the scope and impact of the hospital exemption also remains unclear with interpretation of this aspect of the regulations likely to vary across member states.

Business models and cost-effectiveness

Regulation is not the only area where member state differences means that, with regard to RM, the EU falls short of its goal to create a single market. Whilst the licensing system has been centralised, the EU remains a fragmented health care market with diverse reimbursement systems and varied uptake of new medical technologies. Demonstrating cost-effectiveness and gaining positive decisions from Health Technology Assessment (HTA) bodies is seen as a significant challenge for industry and there was concern that HTA bodies had not begun to address the question of how to evaluate RM products and services. One UK policymaker stated that they had been in discussion with the National Institute for Health and Clinical Excellence (NICE) about this issue and that it was clear that NICE were only just beginning to consider the methodological challenges. In relation to cost-effectiveness some industry interviewees expressed the concern that many of the cost savings that RM products might offer would be outside the health care budget and that current methods of assessment would not take these into account.

Linked to the question of cost-effectiveness was the issue of business models. Many of those in the field express the view that the RM sector has yet to demonstrate the sustainability of business models for producing cell therapies. The cost of developing products, the cost of production, and the size of the markets are all factors which throw doubt on the viability of business models for the sector.

> One is these products are going to be expensive inevitably and so you have to think about the profit margin on these products as well as how big is the market really going to be, how much of the market can you penetrate... when I start now analysing really the rate of market penetration in terms you know the amount the market we are going to be able to capture, I'm still struggling to see how you can develop a viable business model.
>
> (UK interview)

Perhaps unsurprisingly, one industry executive stressed the importance which their company's senior management placed on the sector having some successes which would demonstrate the value of RM, a view echoed by an industry veteran who suggested that what was needed was a genuinely novel application:

> The bottleneck which no one mentions is these things don't work in many cases. They do work in some cases, and they work spectacularly in some cases, but no-one's nailed diabetes; no-one's got a heart repaired... no one has functionally done something yet with a cell that has not been done with something else.
>
> (RM industry executive)

It would, however, be an oversimplification to present the existence of firms producing more mature technologies as being solely due to regulatory and economic dynamics. While most bioscaffolds are aimed at 'classical' TE applications such as repair of cartilage and skeletal tissue, a small but significant number of bioscaffold products in development by European firms highlight an area of potential future expansion for this technology platform. More advanced tissue replacement programs will require the generation of functional three-dimensional cellular structures. While stem cells, especially pluripotent stem cells, may be able to generate all of the tissue types required, they will still need to be grown on three-dimensional bioscaffolds to replicate the complex hierarchical arrangement of naturally occurring human tissues, including, for example, sufficient vascularisation to allow the growing three-dimensional cell culture to survive (Williams and Sebastine, 2005). Thus three-dimensional cell culture is a potential area of future technological development in which the 'mature' technology of bioscaffolds can complement the development of the novel technology platform of pluripotent stem cell therapies. Geels (2002) has noted the importance of this kind of technological 'hybridisation' whereby the development of a highly disruptive novel technology is supported in its earlier stages by linking up with existing technologies to solve particular bottlenecks.

Stem cell therapies

With regard to the sourcing of stem cells, almost all the stem cell therapies being developed by European RM firms involve haematopoietic or mesenchymal lineages – that is 'adult' stem cell types derived from bone marrow tissue. No European firms are developing hESC-based therapies (as opposed to supplying hESC lines for research), most probably because of the combination of the variable state positions on the acceptability of hESC research and the refusal to grant patent protection to inventions which involve the destruction of human embryos by the European Patent Office. This issue of whether hESC therapies will be patentable in the EU was cited by one RM industry executive as an obstacle to investment and again this is an area where it was widely felt that US competitors enjoy a comparative advantage. However, there were divergent views about IPR; for instance, another RM executive suggested that patenting in the RM field is difficult not because of EU blocks on embryonic stem cell-based patents, but because of the amount of prior art. One pharma executive described IPR as 'a minefield' because of the lack of certainty about who owns what and about which IPR is going to be most important, and suggested that the 20-year life of a patent was too short for cell therapies because the much lengthier R&D process left companies insufficient time on the market to recoup their investment before the entry of competitors. The relative importance of alternative forms of IPR such as trade secrets and know-how were emphasised by a number of interviewees, again suggesting a marked difference between the RM sector and the wider biopharmaceutical industry.

Only a small number of European firms, of which ReNeuron (UK) is the most notable, use foetally derived stem cell lines. In terms of visions of a commercially viable RM, European firms have clearly opted in the main for better clinically characterised and less ethically risky stem cell technologies over the greater potential (and greater technical challenges) of hESCs. To date only two European firms have invested in iPS cell technology. Of these the French company Ectycell (a subsidiary of Cellectis) is the most serious contender having licensed proven iPS Cell technology from Japan in 2010. In this, the European RM industry can again be said to lag behind the United States where a number of firms, including Fate Therapeutics (San Diego, CA), iPerian (San Francisco, CA), and Cellular Dynamics Inc. (Madison, Wisconsin) are all developing iPS technology (although primarily as a drug screening and toxicity testing platform). Again, issues of IPR uncertainty are also likely to contribute to caution on the part of investors.

Future trajectories for RM in Europe?

Of all the factors shaping European RM innovation, the current financial climate and the particular challenges it poses for a triple helix/bioeconomy model of investor-supported commercial development of novel disruptive technologies appear the most pressing in the near term. Financial analysis of the global RM industry strongly suggests that even before the Bear market transition of 2008 many RM firms were heavily in debt and failing to produce convincing market signals that they would achieve profitability in anything like the timeframes required by VC investors. Even under optimal financial conditions a high attrition rate is expected among start-up biotech firms. As a 'rule of thumb' calculation, VC firms anticipate that out of ten companies invested in only one will make a profit and another will break even while the remaining eight companies will fail (and in general are liquidated to return whatever capital is left to the VC firm) (Bottazzi and Da Rin, 2002; Talmor and Cuny, 2005). In the current risk-averse financial climate, the presence of severe levels of indebtedness amongst a core of RM firms threatens the lifecycle sustainability of the share-owner-led side to the industry. Furthermore, the model which has enabled publicly listed firms to gain ground in the US and European innovation systems may not have existed outwith of financialisation and the high-risk investment cultures allied to prior Bull market conditions that supported much of the current RM industry.

This in turn has called into question the viability of current business models for RM products and services. From a nation state and EU policy perspective the major issue may now be which sectors of the RM industry to target with strategic support and how best to do this.

This is not an obvious or straightforward choice as from an economic perspective allogeneic stem cell therapies are the most commercially promising option, but also the option with the greatest technical, regulatory and temporal barriers to overcome. Autologous cell therapies are more likely to yield clinical successes in the shorter term, but are also less ambitious in terms of treating chronic/acute illness, are less commercially attractive, and include some TE applications that have only limited markets and cannot support a significant firm base (i.e., there is a limited demand for autologous chondrocyte replacement therapies and little space or incentive for new firms to enter that market). Accordingly, there are a growing number of senior figures in the RM sector who believe that alternative innovation models are required which place greater emphasis on innovation within the public sector. These 'alternative' innovation models tend to involve, to a greater or lesser extent, aspects of the hidden innovation system described by Hicks and Katz (1996) and Hopkins (2006).

One option is to simply delay commercial spin-out, developing a product through Phase I and Phase II trials within an academic setting and then commercialising the R&D operation for Phase III trials. This alternative keeps the same goal of a commercial company producing RM products but requires a purely public investment in the risk of early failure, making the investment more attractive to venture capitalists since there is both lower risk and a more rapid return on their investment. A degree of support for this idea was evident among RM industry interviewees:

> [T]he whole field is moving late stage and for stem cells yes, great [...] it's not mainstream, so it takes time [...] and so my recommendation to many companies is stay in the academic setting the longer you can...because first you will have to waste a lot of your energy fundraising...and then at the end you will probably not make it.
>
> (Spain interview 8, VC executive)

McAllister et al. (2008) have argued that even with the attendant risks of slower development allowing entry of competitors, the lack of investment capital means that 'adding value and mitigating risk under the academic umbrella is an attractive scenario' (2008, p. 928). More recently, in an interview with *Nature Biotechnology*, leading US scientist Irving Weismann, who has multiple experiences of attempting to commercialise RM, evinced a similar sentiment, commenting, 'I wouldn't start a company now unless I had a pretty high degree of control and, much more importantly, had progressed in the university through at least phase 1/2 trials' (Nature Biotechnology, 2011, p. 194). While this model is essentially a (temporal) reworking of the triple helix, it does acknowledge that the risks and uncertainties of developing disruptive, novel, and uncertain biomedical therapies are not necessarily suitable to the requirements of early spin-out commercial development. The likelihood of a 'long-haul' non-linear development pathway, with multiple cycles of research work between the laboratory and the clinic, as exemplified by the case of HSCs, is particularly at odds with the continuous cycles of promise required to attract and maintain financial support in an investment-dependent biotech firm.

Another option, closer in character to the hidden innovation model, is to move cell therapy development more directly into the clinical setting. Weismann also raised this possibility, drawing explicitly on the example of prior attempts at HSC commercialisation:

> For example, if SysTemix had succeeded with its early plan to establish HSC separation units, it would have done this next to a hospital. So why not partner with the hospital to establish and run such units? The hospital and medical school could experiment with how to set up an efficient HSC isolation and transplant and clinical care service, and how to resolve issues of compensation. Should you do it in an outpatient setting? Should you have hospice units? As they explore these issues, I think a model will emerge.
>
> (Nature Biotechnology, 2011, p. 194)

Small-scale hospital-based trials would certainly align with the predominantly autologous cell therapies currently favoured in Europe and there is evidence, in the form of bioengineered blood vessels and tracheas (Adams, 2012), to suggest that successful regenerative treatments, which are unlikely to be commercially viable at present, can be achieved in a hospital setting. Another benefit of the clinical route is the use of hospital exemption rules to allow clinical testing of cell therapies outwith the rigours and expense of full-blown clinical trials. In April 2012, the Texas Medical Board implemented new rules allowing physicians to apply non-approved adult stem cell therapies to consenting patients in the context of medical research (Park, 2012). While the board's decision sets a precedent for expanded research use of stem cells, the decision has been a controversial one and it remains to be seen how the situation in Texas will play out.

Problems also remain in transferring commercially developed RM technologies into clinical settings – one of the reasons that firms are often peripheral to the hidden innovation system is the differences in institutional cultures, practices, and priorities. Clinicians are in a much better position to understand and act within hospital systems than the more business-focused management teams of RM and biotech firms. Conversely, those practices which can successfully mobilise new therapies in a clinical setting often do not meet the standardisation and quality management requirements of a commercially developed product – note Hopkins' (2006) observation on the underregulation of early genetic services developed within the NHS. It is also the case that clinical development may be better suited to autologous cell therapies than allogeneic ones. If viable allogeneic therapies are to emerge it may require innovation in other areas, especially the governance of assessment and evaluation of cell therapies. The difficulties of using standard clinical trials as the 'gold standard' for regulatory decision-making with cell therapies have already been outlined in this chapter. In the recent strategy for RM in the United Kingdom the Medical Research Council stated:

> [I]t is not clear that classic drug trial designs are appropriate for regenerative medicine products, given their higher levels of uncertainty and their likely focus on ultra-orphan conditions. New trial designs may therefore be required, perhaps more adaptive in nature. Such methodological changes could affect trial governance and ethics.
>
> (MRC, 2012, p. 16)

It may also be that data and experience from clinical development of autologous cell therapies is needed to help develop and improve biological standards and clinical practices for the evaluation and testing of allogeneic stem cell therapies in the longer term. Thus the hidden innovation system is not a panacea for getting RM technologies to the clinic, but an option which is due careful consideration in the current circumstances. The increasing integration of commercial cord blood banking in clinical settings (Brown et al., 2011; Machin et al., 2012) may provide a model for RM developers to follow.

Conclusion

The European RM industry is highly heterogeneous and stratified. Extant firms are engaged in pursuing a range of technology options and business models from biodegradable bioscaffolds and autologous somatic cell therapies associated with 'tissue engineering' to stem cell therapies for chronic illnesses. The more established firms tend to be in the 'TE' applications or in the service sector providing reagents and tools (including stem cell lines) to academic and commercial developers. The majority of stem cell therapies remain in the early stages of clinical development.

The difficulties inherent in the dominant bioeconomy model of commercial innovation are heightened in certain ways for European firms; VC in Europe is historically less available, especially for high-risk enterprises, than in the United States. The EU also constitutes a fragmented internal market and, until comparatively recently, has also lacked a harmonised regulatory environment. Ethical and legal restrictions on certain types of stem cell research also vary considerably across member states.

These factors (among others) appear to have favoured the adoption of comparatively risk-averse business models; a strong preference for autologous cell therapies; low uptake of highly controversial and unproved technologies (no hESC products and only recent engagement with iPS); and a preference for better characterised human cell types such as haematopoietic and mesenchymal stem cell therapies. Nonetheless, the current financial climate throws the commercial viability of developing even the most promising cell therapies through a 'traditional' biopharmaceutical innovation model into doubt. New business models, involving greater use of the public sector, are currently being considered as well as further regulatory adaptations to meet the particular material, social, and economic requirements of developing therapies based on living human tissue.

There are reasons for cautious optimism about the prospect of making greater use of hospital-based innovation routes although recourse to the 'hidden innovation system' should not, in itself, be considered a solution for the full range of difficulties besetting RM. Significant issues around IPR remain a concern to the RM industry, although it is worth reiterating that these concerns lie as much with the difficulties in gaining freedom to operate as they do with the much more publicised decision of the European Court of Justine to refuse the granting of patents on inventions that involve the destruction of human embryos, as is currently the case with hESCs.

In the European context autologous stem cell therapies developed within academic and/or clinical settings present the best prospect for achieving clinical successes in the medium term. This may involve novel institutional and governance arrangements such as the formation of in vitro fertilisation (IVF)-style clinics for autologous cell therapy or firm–hospital collaborations akin to the way commercial cord blood banks currently operate. Commercial development of allogeneic stem cell therapies for chronic illnesses will remain a longer-term prospect and may benefit from increasingly successful deployment of autologous therapies only indirectly as greater experience and know-how is built up among a body of clinicians and researchers about how, for example, transplanted stem cells behave within the body and how they can best be tracked. However, as we go on to discuss in the next chapter, the *regulatory* warrant that experience and know-how can claim is no straightforward matter, and can vary considerably at the global level.

Note

1. For the full summary of decisions see http://www.ema.europa.eu/ema/index.jsp?curl=pages/regulation/general/general_content_000301.jsp&mid=WC0b01ac05800862c0

References

Adams, S. (2012) Girl, 10, receives major vein grown from own stem cells, *The Telegraph*, 14 June 2012. Available at http://www.telegraph.co.uk/health/healthnews/9329455/Girl-10-receives-major-vein-grown-from-own-stem-cells.html, accessed 18 June 2012.

Atala, A. (2007) Engineering tissues, organs and cells, *Journal of Tissue Engineering and Regenerative Medicine*, 1: 83–96.

Bahadur, G., M. Morrison and L. Machin (2010) Beyond the 'embryo question': human embryonic stem cell ethics in the context of biomaterial donation in the UK, *Reproductive Biomedicine Online*, 21: 868–874.

Bajada, S., I. Mazakova, J. B. Richardson and N. Ashammakhi (2008) Updates on stem cells and their applications in regenerative medicine, *Journal of Tissue Engineering and Regenerative Medicine*, 2(4): 169–183.

Barber, B. and T. Odean (2001) Boys will be boys: gender, overconfidence and common stock investment, *Quarterly Journal of Economics*, 116(1): 261–292.

Belmonte, J. C. I., J. Ellis, K. Hochedlinger and S. Yamanaka (2009) Induced pluripotent stem cells and reprogramming: seeing the science through the hype, *Nature Reviews: Genetics*, 10(December): 878–883.

Birch, K. (2006) The neoliberal underpinnings of the bioeconomy: the ideological discourses and practices of economic competitiveness, *Genomics, Society and Policy*, 2: 1–15.

Borup, M., N. Brown, K. Konrad and H. Van Lente (2006) The sociology of expectations in science and technology, *Technology Analysis and Strategic Management*, 18: 285–298.

Bottazzi, L. and M. Da Rin (2002) European venture capital, *Economic Policy*, April: 230–269.

Brown, N., A. Kraft and P. Martin (2006) The promissory pasts of blood stem cells, *BioSocieties*, 1(3): 329–348.

Brown, N., L. Machin and C. McLeod (2011) Immunitary bioeconomy: the economization of life in the international cord blood market, *Social Science and Medicine*, 72(7): 1115–1122.

Brown, N., B. Rappert and A. Webster (eds) (2000) *Contested Futures: A Sociology of Prospective Techno-Science*. Aldershot: Ashgate.

Coenen, L., J. Moodysson and B. Asheim (2004) Nodes, networks and proximities: on the knowledge dynamics of the Medicon Valley Biotech Cluster, *European Planning Studies*, 12(7): 1003–1018.

Cooke, P. (2001) New economy innovation systems: biotechnology in Europe and the USA, *Industry and Innovation*, 8: 267–289.

Cooke, P. (2004) Life science clusters and regional science policy, *Urban Studies*, 41(5/6): 1113–1131.

Cooper, M. (2008) *Life as Surplus: Biotechnology & Capitalism in the Neoliberal Era*. Seattle: University of Washington Press.

Daar, A. S. and H. L. Greenwood (2007) A proposed definition of regenerative medicine, *Tissue Engineering*, 1(3): 179–184.

Daley, G. Q. and D. T. Scadden (2008) Prospects for stem cell-based therapy, *Cell*, 132(4): 544–548.

Dumit, J. (2003) A pharmaceutical grammar: Drugs for life and direct-to-consumer advertising in an era of surplus health. Unpublished paper, Department of Anthropology, University of California, Davis.

Ernst and Young (2009) Beyond borders: the global biotechnology report 2009, *Biotechnology Journal*, 4: 1108–1110.

Fairchild, P. J., S. Cartland, K. F. Nolan and H. Waldmann (2004) Embryonic stem cells and the challenge of transplantation tolerance, *Trends in Immunology*, 25(9): 465–470.

Fairchild, P. J., N. J. Robertson, S. L. Minger and H. Waldmann (2007) Embryonic stem cells: protecting pluripotency from alloreactivity, *Current Opinion in Immunology*, 19: 596–602.

Faulkner, A. (2009) Device or drug? Governation of tissue engineering, in A. Faulkner (ed.) *Medical Technology into Healthcare and Society: A Sociology of Devices, Innovation and Governance*. Basingstoke: Palgrave Macmillan, pp. 159–187.

Faulkner, A., I. Geesink, J. Kent and D. Fitzpatrick (2003) Human tissue engineered products – drugs or devices? Tackling the regulatory vacuum, *British Medical Journal*, 326: 1159–1160.

Gardner, R. L. (2007) Stem cells and regenerative medicine: principles, prospects and problems, *Comptes Rendus Biologies*, 330(6–7): 465–473.

Geels, F. W. (2002) Technological transitions as evolutionary reconfiguration processes: a multi-level perspective and a case-study, *Research Policy*, 31(8/9): 1257–1274.

Gruber, A. (2009) Biotech funding trends: insights from entrepreneurs and investors, *Biotechnology Journal*, 4: 1102–1105.

Guertin, P. (2009) The biotechnology industry: what's next? *Biotechnology Journal*, 4: 1124–1131.

Hicks, D. and J. Katz (1996) Hospitals: the hidden research system, *Science and Public Policy*, 23(5): 297–304.

Hirshleifer, D. (2001) Investor psychology and asset pricing, *Journal of Finance*, 56(4): 1533–1597.

Hope, W. (2009) Conflicting temporalities: state, nation, economy and democracy under global capitalism, *Time and Society*, 18: 62–85.

Hopkins, M. M. (2006) The hidden research system: the evolution of cytogenetic testing in the national health service, *Science as Culture*, 15(3): 253–276.

Hu, B-Y., J. P. Weick, J. Yu, L. -X. Ma, X.-Q. Zhang, J. A. Thomson and S. -C. Zhanga (2010) Neural differentiation of human induced pluripotent stem cells follows developmental principles but with variable potency, *PNAS*, 107(9): 4335–4340.

ISSCR (2008) ISSCR guidelines for the clinical translation of stem cells. International Society for Stem Cell Research. Available at http://www.isscr.org// clinical_trans/pdfs/ISSCRGLClinicalTrans.pdf, accessed 19 June 2012.

Kemp, P. (2006) History of regenerative medicine: looking backwards to move forwards, *Regenerative Medicine*, 1(5): 653–669.

Kent, J., A. Faulkner, I. Geesink and D. Fitzpatrick (2006) Culturing cells, reproducing and regulating the self, *Body and Society*, 12(2): 1–23.

Kewell, B. and A. Webster (2009) A tale of 'cautious pessimism': biotechnology, recession and the 'new economy', *Biotechnology Journal*, 4: 1106–1110.

Leydesdorff, L. and H. Etzkowitz (1996) Emergence of a triplehelix of university–industry–government relations, *Science and Public Policy*, 23: 279–286.

Linsley, P. and A. Linsley (2009) Cultural theory of risk and the credit crisis, *Journal of Risk and Governance*, 1: 3.

Lysaght, M., A. Jaklenec and E. Deweerd (2008) Great expectations: private sector activity in tissue engineering, regenerative medicine, and stem cell therapeutics, *Tissue Engineering: Part A*, 14(2): 305–319.

Machin, L., N. Brown and D. McLeod (2012) 'Two's company – Three's a crowd': the collection of umbilical cord blood for commercial stem cell banks in England and the midwifery profession, *Midwifery*, 28(3): 358–365.

Marks and Clerk UK (2009) News item: funding gap will lead to further consolidation of biotech industry, finds new Marks and Clerk Research, May. Available at http://www.marks-clerk.com/uk/attorneys/news/mewsitem.aspx?item=270, accessed 26 May 2009.

Marrus, S. (2009) How to save biotech. The Scientist.com. Available at http:// www.thescientist.com/news/print/55425/, accessed 15 January 2010.

Martin, P., N. Brown and A. Kraft (2008) From Bedside to Bench? Communities of Promise,Translational Research and the Making of Blood Stem Cells. *Science as Culture*, 17(1): 29–41

Martin, P., R. Hawksley and A. Turner (2009) *The Commercial Development of Cell Therapy- Lessons for the Future? Survey of the Cell Therapy Industry and the Main Products in Use and Development. Part 1: Summary of Findings.* EPSRC Remedi Report, Institute for Science and Society, The University of Nottingham.

Martin, P. A., C. Coveny, A. Kraft, N. Brown and P. Bath (2006) The commercial development of stem cell technology: lessons from the past, strategies for the future, *Regenerative Medicine*, 1(6): 801–807.

Mason, C. (2007) Regenerative medicine 2.0, *Regenerative Medicine*, 2(1): 11–18.

McAllister, T. N., N. Dussere, M. Maruszewski and N. L'Heureux (2008) Cell-based therapeutics from an economic perspective: primed for a commercial success or a research sinkhole?, *Regenerative Medicine* 3(6): 925–937.

Medical Research Council (2012) *A Strategy for UK Regenerative Medicine.* Available at http://www.mrc.ac.uk/Utilities/Documentrecord/index.htm?d= MRC008534, accessed 16 May 2012.

Mirowski, P. (2011) *Science Mart: Privatizing American Science*. Cambridge: Harvard University Press.

Morrison, M. and L. Cornips (2012) Exploring the role of dedicated online biotechnology news providers in the innovation economy, *Science Technology and Human Values*, 37(May 2012): 262–285.

Nature Biotechnology (2011) Profile: Irv Weismann, *Nature Biotechnology*, 29(3): 194.

Nerem, R. M. (2010) Regenerative medicine: the emergence of an industry, *Journal of the Royal Society Interface*, 7(53): S771–S775.

OECD (2009) The bioeconomy to 2030 designing a policy agenda: main findings and policy conclusions, *OECD International Futures Project*. Available at www.oecd.org/futures/bioeconomy/2030, accessed 10 October 2009.

Owen-Smith, J., M. Riccaboni, F. Pammolli and W. Powell (2002) A comparison of U.S. and European university–industry relations in the life sciences, *Management Science*, 48(1): 24–43.

Park, M. (2012) Texas board approves rules on use of stem cells, *New York Times*, 13 April 2012. Available at http://www.nytimes.com/2012/04/14/us/new-rules-on-adult-stem-cells-approved-in-texas.html, accessed 23 June 2012.

Pickering, A. (1995) *The Mangle of Practice: Time, Agency and Science*. Chicago: University of Chicago Press.

Plagnol, A. C., E. Rowley, P. Martin and F. Livesey (2009) Industry perceptions of barriers to commercialisation of regenerative medicine products in the UK, *Regenerative Medicine*, 4(4): 549–559.

Rajan, S. K. (2006) *Biocapital: The Constitution of Postgenomic Life*. Durham: Duke University Press.

Redhead, K. (2008) *Personal Finance and Investments – A Behavioural Finance Perspective*. Abingdon: Routledge.

Sheyn, D., O. Mizrahi, S. Benjamin, Z. Gazit, G. Pelled and D. Gazit (2010) Genetically modified cells in regenerative medicine and tissue engineering, *Advanced Drug Delivery Reviews*, 62(7–8): 683–698.

Suarez-Villa, L. (2009) *Technocapitalism: A Critical Perspective on Technological Innovation and Corporatism*. Philadelphia, PA: Temple University Press.

Tait, J. (2007) System interactions in life science innovation, *Technology Analysis and Strategic Management*, 19(3): 257–277.

Takahashi, K., M. Ohnuki, M. Narita, T. Ichisaka, K. Tomoda and S. Yamanaka (2007) Induction of pluripotent stem cells from adult human fibroblasts by defined factors, *Cell*, 131: 861–872.

Talmor, E. and C. J. Cuny (April 2005) The staging of venture capital financing: Milestone vs. Rounds. EFA 2005 Moscow Meetings Paper. Available at SSRN http://ssrn.com/abstract= 487414 or http://dx.doi.org/10.2139/ssrn.487414, accessed 3 June 2009.

Taylor, B. (n.d.) Bull and Bear markets, past and present. Available at https://www.globalfinancialdata.com/articles/bull_and_bear_markets.doc, accessed 10 October 2009.

Tutton, R. (2011) Promising pessimism: reading the futures to be avoided in biotech, *Social Studies of Science*, 41(3): 411–429.

Van Lente, H. (1993) *Promising Technology: The Dynamics of Expectations in Technological Developments*. Enschede, Netherlands: Proefschrift Universiteit Twente.

Van Merkerk, R. and D. Robinson (2006) Characterizing the emergence of a technological field: expectations, agendas and networks in lab-on-a-chip technologies, *Technology Analysis and Strategic Management*, 18(3–4): 411–428.

Waldby, C. and M. Robert (2006) *Tissue Economies: Blood, Organs, and Cell Lines in Late Capitalism*. Durham: Duke University Press.

Webster, A., C. Haddad and C. Waldby (2011) Experimental heterogeneity and standardisation: stem cell products and the clinical trial process, *BioSocieties*, 6(4): 401–419.

Whitaker, M. (2011) Stem cell therapies: to the clinic through the clinic. Keynote presentation at the REMEDiE International Conference Bringing Regenerative Medicine to the Clinic: Trials and Tribulations in Europe and Beyond; 18–19 April, Bilbao, Spain.

Williams, D. J. and I. M. Sebastine (2005) Tissue engineering and regenerative medicine: manufacturing challenges, *IEE Proceedings Nanobiotechnology*, 152(6): 207–210.

Yu, J., M. A. Vodyani, K. Smuga-Otto, J. Antosiewicz-Bourget, J. L. Frane, S. Tian, J. Nie, G. A. Jonsdottir, V. Ruotti, R. Stewart, et al. (2007) Induced pluripotent stem cell lines derived from human somatic cells, *Science*, 318: 1917–1920.

4
Unruly Objects: Novel Innovation Paths, and Their Regulatory Challenge

Christian Haddad, Haidan Chen, and Herbert Gottweis

What is novel [these days] is not uncertainty;
what is novel is a realization that uncertainty is here to stay.

Zygmunt Bauman

Introduction

During the last 15 years, stem cells have become sites of individual and collective human aspirations, where different scientific, medical, and economic visions and desires have been inscribed and contested, leading to multifold disputes and political controversies. Stem cells made headlines in the global mass media, and became a topic of electoral campaigns and heated parliamentary debates. They have brought stakeholders from different fields together into novel regulatory committees and expert bodies.[1] In recent years, the focus of political and policy attention, professional and public concerns, as well as modes of expert deliberation shifted from basic biological research to efforts to apply these insights in clinical therapies, as stem cells moved into clinical research trials and experimental use (Webster et al., 2011). This push towards the clinic has not been a continuous or homogeneous process, but is characterised by striking frictions and heterogeneities in temporal and spatial terms, as well as with regard to practices. In 2010,

The authors want to thank Ingrid Metzler and Paul Just from the Life-Science-Governance (LSG) research platform, University of Vienna, for valuable comments on earlier drafts of this article.

a US biotech company, after years of preclinical development and intense relations with the competent regulatory authorities, announced the first embryonic stem cell therapy for initial human clinical trials (Roberts, 2010), while in China, a network of stem cell labs, tissue banks, researchers, and affiliated clinics reported to have already *treated* several hundreds of patients with their own adult stem cells (Cyranoski, 2009). In Germany, a prestigious team of scientists and clinicians have, fully supported by their peers and by state authorities, incrementally pursued the development of autologous stem cell treatments for a small range of cardiovascular diseases, whereas about 180 kilometres away, a clinical facility deemed 'dubious' has started to provide unproven and costly cell treatments for basically all kinds of paying patients in the same country (Mendick, 2011). This clinic, operating in a legal grey zone, has already been shut down by regulators, as have been clinics in the United States, Mexico, the Netherlands, Bulgaria, and India. The international stem cell community, in unstable strategic alliances with national governments and protective regulatory agencies, has condemned this sort of experimentalism, and took action against 'rogue clinics', domestic and abroad. However, a great deal of suffering patients, rejecting such forms of protective paternalism, endorsed and actively engaged in new forms of medical experimentality (Chen and Gottweis, 2011).

The global landscape of stem cell therapy turns out, therefore, to be a messy and unruly field, marked by disparate medical, technical, legal configurations, as well as heterogeneous and conflicting ethical, political, and economic valuations. Moreover, it constitutes a rapidly evolving area (Webster, 2013, this volume), characterised by its radical social and epistemic uncertainty, fluidity, and unfixity. As novel alleged findings, breakthroughs, as well as new challenges and setbacks surface in real time, the stem cell field, despite diverse and orchestrated attempts to stabilise it, can be perceived as caught in a state of permanent emergence and transformation (Morrison, 2012), that regularly shatters and reverses several transitory closures and settlements.

In this chapter, we explore stem cell therapy as a site through which to understand the dynamics of contemporary social, technological, and economic transformations in the life sciences and accompanying regimes of biopower. Our aim is less to serve an ultimate picture or to systematically map the problems of the field and suggest definite answers and solutions, but rather to engage in a problematisation of governing under conditions of radical (epistemic and organisational) uncertainty and instability. In the following, we will first theoretically discuss this constellation of endemic uncertainty. Then we move on to some regions of Europe, the United States, and China where stem cells have become the topic of various governmental struggles and interventions and bring into analytical focus various crucial features of the interplay of science, technology, and governance in times of radical indeterminacy and increasing social fluidity.

Liquid forms and messy landscapes

During the last decade(s), many authors have tried, empirically and theoretically, to come to terms with the rapid transformation of societies under pressures of globalisation, technological innovation, and the concomitant erosion of long-standing social forms and formations. 'Liquid modernity' (Bauman, 2000), 'world risk society' (Beck, 1999), 'network society' (Castells, 1996), 'societies of control' (Deleuze, 1992), or 'Empire' (Hardt and Negri, 2000) are but a few popular attempts to frame the evasive constellation characterised by the multiplication and decentring of acknowledged points of references, the crumbling of stable social authorities – be it the nation state, the institutions of Science, faith in progress, or identity formation through stable ethnic, national, or religious communities – and the increasing individual and collective disorientation and social disintegration. Moreover, the proliferation of forms and flows of communication, increasing differentiation of social fields and systems, and the multiplication of authorities have created a constellation of complexity which challenges conventional modes of ordering.

The pluralisation and dislocation of stable reference points, emblematically captured in various forms of post-modern thinking, has also opened up the space for post-positivist and interpretative approaches in the social sciences and policy analysis (Fischer and Forester, 1993; Fischer and Gottweis, 2012), and post-foundational political theory more generally (Marchart, 2007). In this chapter we are interested in the theoretical and practical consequences for both policymaking and policy analysis that emerge with the shifting fabric of society. We hold that, on the one hand, increasing complexity and unruliness is thus an empirical and historical fact resulting from a variety of distinct but interrelated events and dynamics. In this sense, governing – in its broadest meaning – is always concerned with finding ways and mechanisms to order and disentangle social phenomena perceived as complex by rationalistic and administrative means. For instance, the concept of risk has helped to come to terms with growing uncertainty, making it calculable, predictable, and hence governable (cf. Dean, 2010). On the other hand, we argue that concepts such as complexity may also be part of intellectual, organisational, and discursive strategies that help – by acknowledging and embracing complexity as an irreducible fact rather than a state to be overcome – making unclear and messy constellations manageable and governable in the first place. No doubt, risk, for instance, is still a powerful force and tool to manage and control various domains of social life, and has by far not become obsolete as a governmental technology or analytical concept.[2] This is especially true in our area of analysis. Our point is, however, that the predominance of risk – both as an organising principle and as an analytical category – has also been decentred and supplemented by other discourses and practices that are shaped by, and shape, today's fluid, messy, unruly order. It is this challenge of governing under conditions of radical uncertainty and complexity, and its practical implications, that we wish to explore in our chapter.

As the literature on new forms of governance indicates, this might be observed, for instance, in the supplementation of rationalist/technocratic by more argumentative/deliberative modes of decision-making and persuasion (Fischer, 2003). In these lines, some scholars even stress that messy social constellations demand messier methodologies that are looser, more comprehensive, adaptive, and more generous (Law, 2004). In an effort to come to terms with the elusive fluctuating reality, social sciences need to reshape and drop the idea of providing, through thorough rational and empirical analysis, stable descriptions, final answers, and clear solutions to social, economic, and technological challenges or problems (Fisher and Gottweis, 2012).

In an effort to make sense of the intricate manoeuvres and struggles involved in the emergence and partial stabilisation of stem cell therapy as a novel experimental site of contemporary biopolitics, we will start by exploring some constellations, events, and dynamics that bring into view the various condensed forces operating in this fluctuating area of biomedicine. We will depart from a consideration of messiness, fluidity, complexity, and unruliness both as methodological ground zero and as a quasi-ontological diagnosis of the social, political, and epistemic horizon of contemporary formations. Under these conditions of 'endemic uncertainty' (Bauman, 2007), *governing* must be understood as the multiple attempts to order, control, and manage a messy, heterogeneous ensemble of objects – and is thus in itself an ensemble of messy, heterogeneous sets of practices that never arrives at full closure, complete control over its object, and ultimate 'success' regarding its objective.[3] It is this aspect of the inevitable incompleteness of any attempts of governing that we will stress and develop in our analysis (cf. Malpas and Wickham, 1995).

Unruly forces in stem cell therapy

Stem cell therapy, as a heterogeneous phenomenon and conflictual field of practice, is brought into being and shaped by a variety of diverse scientific, medical, commercial, and political pressures, demands, and imaginations. There are several forces that shape the boundaries of the field, operating on various geospatial and organisational levels and through varying temporalities. These forces are constituted in complex and elusive relations of interdependence and mutual determination. On the one hand, stem cells have become an important field of attention and a calculated prerogative of many states. National governments have seen an opportunity with stem cells and tissue research to tackle a variety of political and policy challenges, ranging from innovation in health research and healthcare policy to international reputation and effective economic and research strategies in an increasingly global, knowledge-based economy. National governments have become supporters and quasi-entrepreneurs in the field of advanced biomedical research, and much effort is directed towards enabling and promoting stem cell research and therapy to create health, wealth, and technoscientific progress. However, stem cells have also created ethical concerns and political controversies within the state, constituting an embattled field of biomedical research (Geesink et al., 2008; Gottweis et al., 2009). The nation state, as a spatial category to define prerogatives and legal-regulatory zones, is furthermore challenged by emerging stem cell therapies, as, for instance, researchers collaborate in transnational networks which deploy various self-regulatory mechanisms, or patients travel abroad to seek treatments which are prohibited in their home counties. In this highly innovative and complex area of research characterised by its global character, some centres of scientific excellence in selected countries provide the forefront of this branch of science. In this constellation, national regulatory science institutions are likely to struggle to provide and assemble the necessary knowledge and expertise demanded by advanced cell therapies. In Europe, such challenge crucially shaped the legal form of the new regulatory regime governing advanced therapies, which brought into being a newly established supranational scientific committee – assembling the scarce expertise to be found in the member states – in charge of advanced therapies. Governments, in the narrow sense of the executive branch, however, constitute only one powerful force within the contemporary network state which must be perceived less than a single actor, but a complex set of relations and connections that take shape and materialise in a variety of organisations, institutions, and agencies that are interlinked and operate in multidimensional networks, or assemblages.[4] As scholars of the capitalist state might argue, one major struggle that materialises and concentrates in these heterogeneous formations of the state lies in the coexistence of conflicting rationalities both to serve the national industries and attract global capital and to protect its citizen-subjects (Carnoy and Castells, 2001). These relations have already begun to be examined with regard to the global biomedical industries (Thacker, 2005; Sunder Rajan, 2006; Cooper, 2008).

Hence, stem cells appear as contested entities – what elsewhere has been called 'bio-objects' (Vermeulen et al., 2012) – at various sites within the rationalities of the state, its bureaucracies and affiliated, networked agencies. In these constellations, there emerge several conflicts out of the divergent prerogatives and rationalities embodied in the respective agencies. Thus in this specific constellation there exist science and technology policy, predominantly preoccupied with funding and stimulating basic and applied science and technology research, as well as what can be called protective governmental apparatuses, consisting of both public and private organisations, or rather hybrid forms thereof, preoccupied with the protection of citizens – patients, research participants, consumers – from bodily (and economic) harm caused by biomedical research and its products. Regulatory agencies and drug authorities seek to control and monitor clinical research trials, the safety and effectiveness claims of drugs or treatments, as well as market access of new therapies. Stem cells, and cell and tissue products more generally, have in many countries disturbed the proven, established ways of ordering through classification and submission for approval to established regulatory regimes. In the United States, for instance, the Food and Drug Administration (FDA) was struggling in the early 1990s to integrate human cell and tissue products into its well-proven regulatory structure based on the three pillars discriminating between drugs, biologics, and medical devices (Fink, 2009). Also, in the European Union (EU) and in China stem cells have troubled the institutional arrangements of biomedical research and drug regulation in their respective particular ways (Chen, 2009; Faulkner, 2009). As we will explore later in more detail, the institutional arrangements governing biomedical research and clinical trials vary to a more or less tangible degree from country to country. They share, however, some structural characteristics in so far as conventional analytical categories such as regional/global, science/politics or state/industry fail in an effort to render the complexities of stem cell therapy governance graspable. Even more, far from having the capacity to sufficiently control experimental and clinical stem cell use, these regulatory agencies are themselves part of porous, messy governance architectures providing unstable and partial fixations to these fluid practices. However, they form powerful passage points and central nodes in the governance of stem cell therapy.

Usually in these architectures, regulatory agencies are complemented by research ethics committees (RECs), or institutional review boards, located at various administrative levels. Whereas most clinical studies are evaluated and approved by local RECs, some innovative and hence risky approaches are subjected to separate, specialised committees such as the Recombinant DNA Advisory Committee in the United States or the Gene Therapy Advisory Committee (GTAC) in the United Kingdom. The rationale here is that novel approaches such as stem cells demand specific expertise and extra attention because of the high degree of uncertainty, complexity, and lack of experience involved in its application. GTAC, for instance, as its name suggests, was created in the 1990s to supervise gene therapy research. In the last years, however, gene therapy has gradually become less risky in the eyes of regulators due to the advance of scientific knowledge and clinical experience in that field, so that many gene therapy trials no longer need to go through GTAC approval and are evaluated only by local RECs. GTAC, however, has not been decommissioned, but its remit was rather expanded to new uncertain technologies, such as stem cell therapies and other emerging fields of biomedicine. It appears that with rapidly transforming biomedical landscapes, agencies are themselves caught in a state of permanent dislocation and reorientation. The role of ethics review as part of the protective governmental apparatuses has been complicated further by the privatisation and commercialisation of clinical research of the last decades (Mirowski and van Horn, 2005). This development has also led to the emergence of private RECs, often as part of contract research organisations that take over the tasks of public RECs, triggering heated debates over democratic control and legitimacy, and potential conflict of interests (Elliot and Lemmens, 2005; Emanuel et al., 2006). Furthermore, these protective apparatuses are equally exposed to dynamics of globalisation, as many international initiatives and organisations develop and promote global standards regarding both technical aspects of research and its ethical conduct.

Another powerful force shaping the politics of stem cell therapy is the emergence of new forms of empowered patienthood consisting in post-paternalistic models of therapeutic decision-making, direct involvement in political and policy processes, as well as a willingness to more radical forms of experimentality (see Landzelius, 2006; Ganchoff, 2008; Chen and Gottweis, 2011; Langstrupp, 2011). Observations of this phenomenon in the field of experimental stem cell treatments, as well as other forms of promising medicines, suggest that chronically or terminally ill patients with no treatment prospect in their country do not eschew costs and risks involved in experimental treatments. In many cases, it is Western patients travelling around the globe to places where stem cell providers operate in legal grey zones or within less strictly regulated environments, such as Mexico, China, or India. These movements pose challenges to governments and their protective apparatuses as well as to the stem cell community and its associated industries: 'quacks' and fraudulent doctors, in their perspective, are a threat to respectable scientists and the 'ethical industry' (Mason and Manzotti, 2010), who have invested in the creation and promotion of both ethical and commercial values (Salter and Salter, 2007; Sunder Rajan, 2006). 'Rogue' stem cell providers potentially shake the young field that emerges under many pressures and struggles. Unable to prevent or directly punish offshore clinics, regulators and the protective governmental apparatuses seek to counteract these dynamics by educational and informational programs (such as the 'Patient Handbook' of the International Society of Stem Cell Research, see ISSCR, 2008), international initiatives (such as the BIONET coordination action between the EU and China, BIONET, 2010), and legal sanctions, where possible. It needs to be mentioned that these clinics are not always found in 'non-Western' countries and emerging economies, but sometimes in the very heart of advanced industrialised societies: there are many cases from the United States, Germany, or the Netherlands where experimental stem cell treatments were provided to paying patients under ethically questionable conditions, using legal loop holes and regulatory grey zones (see Wahlberg and Streitfellner, 2009).

Moreover, the picture here is, again, fractured: the many reported cases of such experimental stem cell clinics indicate that such operators act out of a variety of motives ranging from intentional, fraudulent exploitation of desperate patients to doctors acting in their best knowledge and judgment in an optimistic effort to cure the terminally ill, and businesses developing and testing alternative models of clinical translation and product development in the absence of clear legal and regulatory frameworks. As we will see, the combined strategies of various stakeholders in the field to create an ethical – opposed to an unethical – biomedical industry is not merely 'ethical' in itself, but part of 'integral' business strategies consisting in the creation and allocation of symbolic ('ethical') capital and its concomitant delegitimising effects on potential competitors (see Franklin, 2003; Sunder Rajan, 2006 on the relations of ethics and biocapital).[5]

Boundary irritations and demanding integrations: stem cells at the regulatory science agencies

In the 1990s, regulators at the US FDA came under pressure, as cell and tissue-engineered products were seized by the 'regulatory gaze' of the agency. Biotech company Genzyme had put its tissue product for knee repair, Carticel, on the market. At that time it was unclear with respect to the existing regulatory frameworks how to regulate such new tissue and cell products. Containing living human cells, it was suggested that such products which usually fall under the medical device regulation need to be reassessed and reframed due to their biological properties.[6] Carticel was granted a conditional manufacturing and commercialising license as an unproven therapy, while policymakers and FDA officials, in an awareness of the highly fragmented regulatory situation with its gaps, overlaps, and indeterminacies, were working on a new set of regulations able to embrace such novel products from the field of regenerative medicine (Anon., 1997). For many decades, the FDA's approach to regulating novel medical products has been grounded in a three pillar architecture that differentiates basically between drugs, biologics, and devices. These pillars correspond with varying historically shaped organisational and institutional structures and varying regulatory priorities, and these pillars are reified by a consensus of those routines, expertise, and regulatory science demonstrations which are seen as necessary to regulate the respective class of products. The FDA, enjoying a considerable reputation[7] among both the regulated firms and the public (Carpenter, 2010), has built its standing on the very modern proposition of continuity and a science-based tradition (Daemmrich and Radin, 2007). The Biologics Control Act, for instance, establishing federal regulations of toxins, viruses, and sera, dates back over a hundred years to 1902. Incorporated into the 1944 Public Health Safety Act, and amended several times since, it still provides the legal foundation of today's regulation of basic and advanced biopharmaceutical products (Halme and Kessler, 2006; Wittlesey et al., 2011).

From the mid 1990s, the FDA was involved in a decade-long process, namely, part of the 'Reinventing Government' initiative that sought to integrate the new regenerative objects into existing frameworks (FDA, 1997). During this period, the 'FDA Modernization Act' was passed, introducing a series of substantial reforms 'to make the FDA ready for the 21st century', from cutting red tape, lowering the regulatory burden for industry, and establishing new regimes of experimental access to developmental drugs (Office of the Press Secretary, 1997). The outcomes of this process were specific regulations for so-called human cell and tissue-based products, which were finally implemented in 2005. What was at stake was not only the difficult categorisation of transboundary objects – combining features of drugs, devices and biologics in yet unacquainted settings – but also the regulatory science innovation demanded by biomedical innovation: the development and validation of appropriate tests that assure the safety, potency, and efficacy of these novel applications. Such efforts are difficult socio-technical projects, involving not only scientific, but also political judgements and calibrations. Regulatory science regimes have to deal with the circumstances that there are hardly any clinical experiences and only few animal models available for stem cell therapies, whereas the latter count as highly problematic 'human proxies' in regenerative medicine (Baker, 2009). Testing the drug–body interactions of chemicals in animals with the intention to draw inferences for human pharmacology may be a challenging task, but its complexities multiply with cell-based therapeutics. One key struggle relates to the contested category of sufficient evidence required to proceed from animal to human trials. Whereas good basic scientists demand an adequate knowledge of the application's mechanism of action (i.e., how the therapy works in the recipient's body), clinicians emphasise only reliable evidence of safety as necessary to move the experiments to the clinic. As one clinical researcher has said,

> Making a mouse ES [Embryonic Stem] cell and putting it into a mouse is probably not going to answer the question of putting human cells into a human. [...] The only place you're going to be able to do the definite experiment is in patients. Therefore it is critical to balance the risk to the patient with the possible benefit (Carpenter quoted in Baker, 2009).

This struggle brings into view the specific inter- or even transdisciplinary character of stem cell therapy research where different, rather incommensurable traditions of research and socio-scientific validation meet. Hence, the struggles over the best practices are not merely technical, scientific, or ethical, but also proxy struggles over the authority and legitimacy of, for instance, basic versus clinical research or ways of knowing established in molecular biology versus in experimental clinical medicine. Regulators and regulatory scientists – including RECs, which are often predominantly staffed with 'bench scientists', as one of our interview partners complained about the situation in the United Kingdom – are, in this respect, predominantly concerned with safety and scientific rigour, which provides one crucial element of the normative basis for an ethical human clinical trial to get started in the first place (Emanuel et al., 2000). Another complicating factor lies in the fact that stem cell products often require demanding manufacturing processes that are challenging both from a technical and from a regulatory point of view. Similar to biopharmaceuticals, the product is dependent on its generative manufacturing process, involving a range of technologies to control and stabilise the entity (Webster et al., 2011). The fact that cell products consist of *living* materials complicates this further.

The regulation of regenerative and cell therapies is frequently portrayed as *intrinsically* complex, as one emblematic problematisation exemplifies. At a European Medicines Agency (EMA) workshop in May 2010 addressing regulatory challenges with stem cell-based therapies, a member of the Finish Medicines Agency and the EMA's new Committee for Advanced Therapies (CAT) demonstrated the degree of complexity involved in stem cell therapies in her presentation. On one of her slides, she showed three images from different regulatory science objects: first, the chemical drug Aspirin represented as a small molecule in its simple, two-dimensional molecule structure, followed by Filgrastim, a large molecule peptide hormone in its three-dimensional structure, which looks much more complex but still intelligibly structured, and finally the very messy picture of a eukaryotic cell, cut in half to show its messy intracellular structures that evokes a rhizomatic assembly rather than an intelligible structured entity (see EMA, 2010; Salmikangas, 2010). Of course, the message of these comparative illustrations lay in an assertion of the complex nature of cell therapies, signalling the intricate ways of knowing and managing such objects. This anecdote might tell a variety of things,[8] but one is of crucial importance here: in this demonstration, regenerative medicine and stem cell therapies are framed as messy objects and the regulatory challenges they bring about are first and foremost owed to their natural, *intrinsic* complexity – a complexity that needs to be tamed, managed and eventually overcome by effective regulatory arrangements based on scientific assessment and expert deliberation. Moreover, in the context of this EMA workshop, gathering regulators, scientists, and stem cell developers, this reference to intrinsic complexity might also be understood as a strategy of establishing common ground and an awareness of the necessity to collaborate across disciplines in an effort to overcome these shared obstacles. In this matter, the reference to the history of biomedicine, or rather a very linear and progressive depiction thereof, serves as a strategic horizon for the community to frame problems and work on their solutions. In an evening event organised by the London Regenerative Medicine Network, a well-known clinical researcher, again, drew on complexity as a rationale for strategic regulatory considerations. In his keynote address, problematising the state of the art of clinical trials in both the gene therapy and the stem cell therapy field, he made the following point: genetics basically is a rather simple, mechanic matter. Therefore one must rigorously study the mechanism of action of the experimental gene, before going into human trials. Stem cell biology, on the other hand, is such complex and intractable matter, that an effort to understand the mechanism of action of a cell therapy would delay the clinical evaluation of such therapies for years. Hence, one should adopt a clinical rationality and go into human clinical trials as soon as the safety of the cells is established to see if the treatment works (cf. Martin, 2009). With regard to the case of the first embryonic stem cell-derived product ever authorised by the FDA for human clinical trials – developed by Geron Corporation, which we will discuss in the subsequent section – this comment might be borne out for the sponsor company submitted a 22,000 page dossier in its clinical trial application, illustrating precisely what complexity and endemic uncertainty does to institutions not prepared to embrace radical complexity.

Of course, a reference to intrinsic complexity – as well as any approach trying to separate intrinsic from external complexities – falls short on understanding the very messy, and indeed complex, constellation of stem cell use and its governance. Governance, even in a narrow meaning of the concept, is always concerned and implies attempts of complexity reduction in an effort to enable interorganisational communication and steering between highly differentiated and virtually incommensurable social systems (Jessop, 1998). Overcoming complexity is indeed a highly valued shared objective among most stakeholders in the stem cell therapy field; however, these processes must be understood as difficult negotiations between conflicting views and perceptions. In such fluid, constantly revolving fields such as the biosciences, overcoming complexity might be an objective that demands for pragmatic solutions and some sort of what could be called 'epistemic diplomacy' to settle conflicting views and discourses in a transdisciplinary field of science. Similarly, Linda Hogle has described the processes by which tissue-engineered products were made governable through the establishment of a 'pragmatic objectivity' (Hogle, 2009), able to settle different challenges deriving from scientific uncertainty, regulatory requirements, and pressures from and dissent among medical, bioindustrial, and policy constituencies.[9] Such pragmatic decisions might conflict with the very modern idea of a thoroughly rationalist, science-based regulatory agency such as the FDA, but appear to gain momentum in today's liquid constellations.

Fragmented European tissues

Similar obstacles have arisen for policymakers, regulators, and drug authorities in the EU. It was not so clear how to deal with these novel and hybrid technologies that blur the well-defined zones of laboratory research, bioengineering, surgery, and drug development (Brown et al., 2006; Kent et al., 2006). At first, gene and somatic cell therapies challenged the regime of harmonised European medicines regulations. Eventually, the policy processes led to the creation of a unified framework, subsuming all innovative 'advanced therapies' as medicinal products under the same umbrella.[10] Studying these policy processes brings into view the complex, mutual permeating forces shaping the European technopolitics that are integral for the governance of technological societies (Barry, 2001).

Since the late 1990s, European policymakers have been engaged in various programs and initiatives to reconfigure European medicines' laws and the regulation of clinical research, development, and drug marketing. From the perspective of the pharmaceutical industries and drug developers, Europe appeared as a fundamentally fragmented regulatory space and market, with heterogeneous laws governing production and marketing of drugs, different competent authorities, and health systems. Nevertheless, Europe has seen increased efforts of harmonisation within the last decade, ranging from clinical trials to harmonised cell and tissue practices and the reorganisation of institutional architectures at the supranational level. Despite these efforts of smoothing the uneven and interrupted European pharmaceutical zone, the drug development process remains distinctively discontinuous, for various reasons: Clinical trials authorisation and supervision – from early Phase I up to pivotal phase III trials – remains in the hands of national competent authorities. In Europe, the clinical development process can, and often does, take place in various EU member states and hence single trials are reviewed, authorised, and evaluated by different regulatory agencies and RECs. All the dispersed studies and collected data relevant for regulatory review are assembled not until the very filing of a marketing authorisation application.

Since 2004, all biologics and otherwise innovative medicines – hence including all regenerative/cell-based therapies – need to go through the centralised procedure of the EMA, and thus Advanced Therapies receive a single European market license, once approved. Whereas this European regime has grown historically, reflecting a series of prominent political struggles, such as the distribution of competences and powers between the national and the supranational level, and might follow a political rationality of European integration that goes far beyond stem cells and even pharmaceuticals, one point is illustrative for the problem of complexity and fluidity involved in knowing and governing stem cell therapies. 'Advanced therapies' were constructed as an umbrella concept to embrace, casually phrased, basically all what is new, risky, and unknown in biomedicine, establishing a new expert body preoccupied with its evaluation and regulation. This policy choice implies that the existing committee in charge for human medicines is not capable of handling such innovative products – and neither are national competent authorities. These advanced therapies have such a disruptive potential, in an epistemic as well as in a political sense, that they demand extra attention. The newly established CAT brings together the scarce expertise from all corners of Europe in an effort to tame and control these unruly entities. The new regime with its new scientific expert committee and its mandatory centralised procedure at the supranational level points to the co-construction of both radically new, 'advanced' scientific and regulatory entities that, although constantly in flux, have come to stay in contemporary political imaginations and rationalities.

We have now explored how stem cells have entered regulatory science institutions and how regulators and the broader protective governmental apparatuses have dealt with these radically evasive, unruly objects. We will now turn out attention to the stem cell industry and clinical research organisations trying to develop stem cell-based therapies and bring them to application via clinical research and/or treatment trials.

Operating in ongoing emerging structures: innovating firms and their business strategies

Probably the most visible company, emblematic for stem cell therapy development and its intricacies, is the Geron Corporation, a biotech company located in Menlo Park, California, that has been involved in the development of clinical-grade nerve cells, derived from one of the early registered human embryonic stem cell lines from the Wisconsin University research lab. Founded in 1990, the company focuses on applied research in the areas of cancer therapies and regenerative medicine. Among its stem cell products in various development stages, it has become publicly visible through its most advanced – and fiercely debated – stem cell-derived product for the treatment of spinal cord injuries that, after years of preclinical development and regulatory scrutiny, was approved for a clinical human safety trial in 2010.

This arduous process is a story of a drudgery, that resulted in various delays, regulatory re-examinations, holds, and, eventually, approval. In this process that took the company about a decade, Geron accumulated a 22,000-page dossier submitted to FDA review (Pollack, 2010). Many observers of the field did not believe that Geron would ever make it through the regulatory process, for various reasons: the approach would be too complicated; it would be too early for stem cell therapies; the product would be too expensive and time-consuming to develop; it would be too difficult to guarantee the safety of the product and to establish valid testing criteria; and so on. But Geron kept on pushing the product further towards clinical trials, and eventually succeeded, as the FDA gave its final favourable decision in mid 2010. The product entering clinical trials constitutes what a regulator scientist in an interview called a 'horror scenario': a highly manipulated cell line derived from an embryonic stem cell line years ago and since cultured and manufactured is introduced in the course of a highly risky surgery through a special device – that needed separate regulatory approval – into the broken spinal cord of a young adult. Given this, the company's decision in November 2011 to shut down its stem cell program was most astonishing for a great number of observers. Even further, the justification for the withdrawal angered patients and patient organisations, who felt abandoned in their hopes for a cure: it was not the technical or the political-regulatory challenges that brought an end to Geron's stem cell programme, but simply a decision made by the management board in an effort to efficiently allocate investments and to reshape the company's business strategy towards their better established cancer therapy programs (Pollack, 2011). The case reflects one specific imaginary of how research, innovation and product development works and should work in contemporary regulatory relations, which regard themselves as best practice through ethical governance. As Science and Technology Studies (STS)-inspired scholars of science policy and innovation have argued, contemporary relations between science, technology, and society are governed by a tacit rationality that innovation can be steered and promoted by politics and policymaking, and that innovation, to be 'successful', depends on proper regimes of regulation. Alex Faulkner, for instance, has termed these productive relations between governance and innovation 'governation' (Faulkner, 2009). Geron indeed was a test case both for regulators and for the entire stem cell industry, which was closely following the developments. In this regard, it was both a positive and a negative test case: positive, in an effort to try out, develop, and refine the ideas and practices to regulate stem cell therapy development; and negative, because stem cells have been perceived as highly risky entities, and great parts of the community did not preclude that introducing these cells to a living human organism might result in completely unexpected, even disastrous effects. Geron was always portrayed as a compliant company that engages in public dialogue and regulatory discussion, working together with FDA to overcome these intrinsic challenges of stem cell products in the clinical trials process.

The case of Geron's stem cell decision, however, subverts and frustrates, to a certain degree, this imagination: the product seemed to work (or was at least promising); the regulatory hurdles were burdensome, but were eventually overcome by well-arranged collaboration between the company and the regulators; nevertheless, this aspiration experienced a sudden and unanticipated end. Today's endemic uncertainty originates from various, sometimes indeterminate, incalculable forces and dynamics, and it is not so clear that good science plus good regulation necessarily leads to the intended outcomes. It could be said that Geron was constructed as a case and symbol representing the dominant operational and normative paradigm of stem cell therapy research and innovation. First worries that Geron's retreat from stem cell therapy might send shockwaves through the broader community, and would negatively impact on biotech and regenerative medicine stocks did not materialise immediately after the event. In a critical acclaim, UK-based Chris Mason – one of the most dedicated spokesmen and promoters of the regenerative medicine industries and chief editor of the industry's journal *Regenerative Medicine* – tried to pour oil over troubled waters and has qualified Geron's symbolic role for the larger community, and announced a 'post-Geron era' (Brindley and Mason, 2012). In line with a series of other articles and comments, the authors suggest that despite its significant, defining role, Geron's withdrawal will not bring about the death blow of stem cell therapy field, as a series of companies are ready to move up and to step out of the shadow of Geron.[11]

However, there is apparently also an entirely different world of stem cell industries that pursue and develop different regulatory and business strategies to bring stem cells to clinical research and use. Many of them challenge the dominant regulatory discourse on how innovating companies should pursue their research and product development. The strategic dichotomisation between an 'ethical industry', which is framed as responsible and transparent, and is compliant with evidence-based scientific and regulatory standards, and an 'unethical industry', consisting on 'rogue clinics' and fraudulent providers, fails to capture the messy worlds of the stem cell industries.[12] We argue that stem cell therapy and its innovating industries are a good area to study the multiple, messy, and indeed unpredictable manoeuvres involved in innovation processes. In their attempts to navigate in the patchy, unclear, and fluid epistemic, technical, and regulatory jungle such companies are always involved in technoscientific, business, and regulatory innovation. In an effort to do so, we will now go into two cases – one from Germany, one from China – that bring into view a crucially varying set of relations and practices to engage with and bring forward stem cell therapies. In exploring these sites in detail we will be able to elaborate further on the complexities of governing unruly cells and their fluid structures.

Innovation from the clinic?

In 2001, clinical researchers in Germany conducted the first clinical trials with patients that have suffered from an acute myocardial infarction, using the patient's own stem cells as therapeutic tissue (Strauer et al., 2002). Such an approach introduced a kind of paradigm shift within cardiovascular medicine, as regeneration of the heart muscle was hitherto thought impossible. Since then, the clinical research field of cardiovascular cell therapy has grown rapidly: many small clinical studies are performed all over Europe, and with some delay in the United States, establishing safety and mild efficacy data of these autologous stem cell treatments (Wilson-Kovacs et al., 2010; Webster et al., 2011). As a result, the European Society for Cardiology (ESC) had already stated in 2006 that there is no more need to conduct small safety studies, but to move forward to larger randomised, controlled trials in order to generate robust efficacy data (Bartunek et al., 2006; see also Martin, 2005). This step, however, confronted the so far rapidly evolving field with serious challenges. As a result, the ESC has created the Task Force for Stem Cells and the Heart in an effort to bring clinical researchers from various European centres together to discuss challenges and harmonise research protocols, and to engage in early discussion with the EMA. In the following we will focus on a research group based in Germany that has sought to establish cardiovascular stem cell therapy clinically and to bring it to market.

At the time when German researchers started to clinically test bone marrow stem cells for heart diseases, German federal regulations were categorising such treatments as tissue transplants, which were individually manufactured and applied to the patient shortly after retrieval of his/her cells. In the eye of the regulator, such a procedure does not fall into the category of drugs or 'medicinal products', hence they were not required to pass through the formal evaluation procedure by randomised controlled clinical trials. The requirements were only a manufacturing licence, controlling and guaranteeing good manufacturing and laboratory procedures (hygiene, quality, etc.). Moreover, the procedure was part of treatment trials within the prerogative of the physician that forms an integral part of clinical practice that is worth its name.[13] In the course of the emergence of the European regulatory framework for 'advanced therapies', such cell therapies were, however, subjected to regimes similar to conventional drugs – which implicated the obligation to run formal clinical trials before a product is authorised. The European Regulation even provides for ex post clinical evaluation of cell and tissue products that are already in clinical use. This transition period and the legal grey zones that they created opened up the space for commercial stem cell providers, operating not in an effort to conduct clinical research and product development, but to commercialise unproven stem cell therapies for paying patients. Meanwhile, the German federal authorities have shut down such a centre, which, nevertheless, operated from 2007 for four years, treating over 3500 patients.

The German cardiology research team from our case study, eager to see their product acquire the status of a licensed therapy, already having 'successfully' completed Phase II trials, were planning to conduct a large Phase III trial anyway. A trial protocol was already approved by the German regulatory authorities in 2006, years before the EU Advanced Therapies Regulation came into force, demanding that by the end of the transition period of three years, all cell therapies that were in research and/or treatment use at that time must have completed clinical trials. For the German research team the crucial question was simply to raise funding and get the trial started. This task turned out to be impossible, even though the product was well advanced and clinically promising, and the estimated costs for the Phase III trial were, for various reasons, significantly lower as in conventional trials within cardiology.[14]

As it became clear that the small company would not succeed in acquiring the funding needed to run the Phase III trial for acute myocardial infarction, the researchers rethought their strategy. It might be perceived as a scientific serendipity that allowed the company to switch the indication of their developmental product from heart attack to a rare vascular disease: the mechanism of action that underlies the treatment – consisting in the regeneration of vessels in ischemic tissues – works in the heart as well as in the limb vessels diseased by a rare type of vasculitis. So the team decided to put the development process for heart repair on hold and to continue to bring the very same product towards market approval – however, for a very different medical condition. This change in indication has substantial effects on the clinical trial: as a rare disease, the company can draw on several regulatory incentives and support mechanisms of the EMA established by the EU to foster and advance rare disease research. These include reduced fees, tighter cooperation with regulators, and – maybe most importantly – an exclusive marketing licence for ten years in Europe and seven years in the United States. This is crucial especially for this company developing a promising, however not patentable product for acquiring funding and for commercialisation after approval. But most significantly, a rare disease designation has an impact on the trial size. As the CEO of the company explained, it is hardly possible to find 1200 eligible trials participants for a disease with only about 10,000 patients in Europe suffering from that condition. In this case, the number of trial subjects required by regulators drops from about 14,000 to about 200, which in turn significantly reduces trial costs and hence increases funding opportunities.

All this illustrates that developers in the fluid regimes of stem cell therapy have also to be creative and experimental in their business strategies that seem to be rather adaptive and fluid themselves. Bringing a stem cell therapy product to clinical trials and, eventually, to the market seems to demand various detours, flexible and even fluctuating scientific and regulatory approaches, as well as some sort of improvisation and an experimental attitude towards innovation models and business strategies. We will now turn our focus to the role of new (as well as traditional) forms of patienthood, as well as the role of the patient as research participant, trial subject, and consumer of experimental drugs and unproven therapies.

China: Governing stem cell science by heterogeneity

Whereas in Europe and the United States government agencies have attempted to deal with the highly heterogeneous constellation of stem cell science by creating homogenous frameworks of regulation, the case of China shows how the messiness of stem cell science can give rise to strategies that incorporate the simultaneous operation of contradictory regulatory strategies as a means to navigate a nation's stem cell research system through the deep waters of global R&D competition.

On the 30th anniversary of Shenzhen Special Economic Zone, China's premier Wen Jiabao visited Shenzhen Beike Biotechnology (hereafter, named Beike), a Chinese biotechnology company dedicated to the development and commercialisation of adult stem cell therapies, and said

> Shenzhen is a new city. The stem cell research that Beike has been doing is frontier biotechnology, and the most promising field to exceed Western developed countries. I hope you could take on heavy responsibilities to turn hope to reality, and to make a country, a nation, as you have expected, lead to be in front in the big field of science. In the future, those in the leading position of stem cell field are neither in Cambridge, in Singapore, nor in America, but in Shenzhen, in your hands, in your brain, in your fighting spirit and your wisdom.
>
> (Ceosz, 2011)

This statement to some commentators came as a surprise, as over the years Beike has drawn a lot of national and international attention for its stem cell treatments on a large scale without performing clinical trials as such (Lau et al., 2008; Kiatpongsan and Sipp, 2009; Wahlberg and Streitfellner, 2009; Chen and Gottweis, 2011). After Sean Hu, the CEO and chairman of Beike took his PhD at the Department of Biochemistry and Biophysics at Gothenburg University in Sweden in 1998, and post-doctoral research in the Department of Biochemistry and Molecular Biology at the University of British Columbia, he moved back to China in 2001 to develop and foster basic and applied stem cell research. He collaborated with doctors in Zhengzhou, Henan Province, who proceeded to treat the first ALS (Amyotrophic Lateral Sclerosis, also known as Motor Neuron Disease) patient in 2001. From 2001 to 2005, over 200 patients with incurable diseases were treated in various studies sponsored by Hu. Once Hu felt comfortable with stem cell treatments, the company was set up on 18 July 2005, with capital from Peking University, Hong Kong University of Science and Technology, and Shenzhen City Hall. Less than one month later after the company was founded, a nine-year-old girl with chronic Guillain–Barre syndrome became the first patient to receive Beike's stem cell treatment.

Beike's business model is to build a platform linking scientists, doctors, research institutes, hospitals, and the company to perform translational stem cell research and clinical application. The company collaborates with doctors and hospitals, provides them with their stem cell technology and equipments, and arranges patients there to receive stem cell treatments. Meanwhile, the company sponsors researchers to conduct basic stem cell research and clinical studies. On Beike's Chinese website (www.beike.cc), the company listed some major events in the process of its development since its founding. On 1 September 2005, 463 Hospital's Cell Treatment and Rehabilitation Centre of the Chinese People's Liberation Army (PLA) became Beike's first collaborative stem cell treatment centre, located in Shenyang, Northeast China. Since then, Beike has expanded to collaborate with both domestic and foreign partners. In November 2007, Beike signed a joint agreement with the Shenzhen Graduate School of Tsinghua University to do stem cell research. In the following year, Beike and China Medical City agreed to build a stem cell industrialisation base at Taizhou where the China Medical city is located. From 2009 to 2011, Beike initiated collaboration with Yokohama-based Biomaster Inc., Bangkok-based SiriCell Technologies Inc., and US-based ThermoGenesis Corp. for mutual development in the stem cell business, and opened the first Chinese rehabilitation centre in Campina, Romania.

The company claimed to have treated patients from many other countries such as the United Kingdom, America, Australia, Canada, and some European countries. In order to attract more patients it set up websites specifically for patients in China, Europe, and the United States, where they can find the information about the company, patients' treatment experience and testimonials, the treatment procedures at Beike, the hospitals where treatments are provided, treatment costs, the nature of the stem cells used for the treatment, how patients could travel to China for treatment, and so on. Patients set up blogs to share their life and their experiences with stem cell treatments in China, which helped to connect them with their families, friends, and other patients. Patients were also active in joining online patient communities such as stem cell awareness association, yahoo health groups, QQ groups (Chen and Gottweis, 2011). It is difficult to assess how many Chinese and foreign patients, respectively, have been treated at Beike's collaborative centres but evidence indicates that they have treated more than 5000 patients with stem cells from Beike's founding in 2005 until 2009 (Zeng, 2009). The number seems to be growing all the time, as the latest patient treatment experience was updated on 6 March 2012 at China Stem Cell News (www.stemcellschina.com), a website organised by Beike to provide up-to-date and on-target information about adult stem cells and current treatments available in China.

Beike has been broadly debated as an indication that China is a centre for 'stem cell tourism' where patients receive unproven therapies in clinics which capitalise on the often desperate situation of patients seeking treatments at all costs (Baker, 2008; Lindvall and Hyun, 2009; MacReady, 2009). In this perspective, China seems to have become the place where no strict regulations are in place to govern stem cell therapies, and scientists and doctors can freely experiment with unproven stem cell techniques. But the picture in China is, indeed, more complex. The fact is that different approaches to translational stem cell research are conducted simultaneously in China. Many scientists follow strict international standards and regulations. In December 2004, the first Chinese stem cell-based therapy 'Bone Marrow-derived Mesenchymal Stem Cell' received official approval from the State Food and Drug Administration (SFDA) to begin Phase I clinical trials, and the application conformed to the Provisions on Drug Registration. This research was conducted by Chunhua Zhao's team from the Centre of Excellence in Tissue Engineering, Chinese Academy of Medical Sciences, and Peking Union Medical College, Beijing. It was the first case in China to receive approval from the SFDA for allogeneic bone marrow-derived mesenchymal stem cells clinical trials on graft-versus-host disease (GVHD). The stem cell-based therapy developed by the Zhao team is similar to Prochymal, which is developed by Osiris Therapeutics, Inc., an American company, and is the only stem cell therapeutic currently designated by the FDA to treat GVHD as both an Orphan Drug and a Fast Track product. It was reported that Prochymal had begun Phase III clinical trials (Osiris, 2012). All the way through its research, Zhao's team had attempted to adopt a model closely in line with the recently developed international standards in stem cell clinical research, in particular the FDA model. China's current heterogeneity of stem cell therapies is largely due to the absence of clear rules and regulations, and, most importantly, the implementation of regulations.

Heterogeneous regulatory constellation

As we have seen so far, both America and Europe have had difficulty in coping with the complexity of stem cell therapies. This was also true in China. In 2003, when the Zhao team submitted their documents to SFDA for review to be approved for clinical trials, the National Institute for the Control of Pharmaceutical and Biological Products (NICPBP), the drug testing institute designated by SFDA, was unsure how to review the therapy and how to set the standards for it. Because this stem cell therapy is allogeneic, the SFDA finally decided that it should belong to the investigational new drug category of biologics, according to the Provisions of Drug Regulation. This regulation is similar to the FDA's position on stem cell-based products. However, in the first review processes of 2005 and 2006, the NICPBP encountered problems in following the Provisions of Drug Regulation to regulate these stem cell-based products and therapies as drugs since the Ministry of Health (MOH) proposed that stem cell therapies should be treated as a medical technology, under the regulation relating to clinical applications of medical technology. The long discussions and negotiations between these two regulatory agencies created grey areas for the stem cell field. Different narratives about the right road to secure the safety, efficacy, and ethics of applied stem cell research competed with each other. This has meant that some researchers and companies have been moving very quickly into this new field of stem cell research, while others have been much more cautious in navigating in this complex regulatory environment. Currently two main different models for innovation with stem cell-based therapies coexist, one with researchers that attempts to follow the SFDA's strict regulation on drug approval requirements and procedures, which is comparable to the FDA model (the Zhao case). They try to demonstrate efficacy and risk assessment in animal models first; only after their preclinical data was peer-reviewed could they start with clinical trials to get their stem cell therapies approved. The other is Beike's model that deploys stem cell treatments with large numbers of patients with the expectation that efficacy and clinical evidence will emerge over time and so enable them to obtain legitimacy in the stem cell field. Currently, the SFDA and MOH are negotiating how to regulate these approaches. The Shanghai group led by Chingli Hu has proposed to differentiate between preclinical research, clinical trials, and clinical applications in the field of stem cell research and application, stressing the differences between experimental stem cell treatments offered to large numbers of patients and stem cell-based treatment trials with a limited numbers of seriously ill patients in the context of a cautious medical innovation model. How this is resolved is still unclear but it is apparent that the Zhao and Beike cases reflect heterogeneous innovation practices and an improvised, heterogeneous regulatory approach.

Conclusions: towards experimental regimes of governance?

Stem cell therapy has indeed shown as an unruly biomedical field of heterogeneous practices that appears as caught in constant flux and reconfiguration. In this chapter we have looked at a variety of sites that have helped to bring into view the complex, elusive setting of emerging stem cell therapies and its governance. Concepts such as messiness, fluidity, and complexity have helped to bring into view some distinct, but interrelated features and forces at work in the heterogeneous boundaries of biomedical research, commonly perceived as radically new or innovative.

We conclude by reflecting on what could be learned from our exploration of stem cell therapies as fluid sites of regulatory science struggles. Hence, our research suggests that governing in such messy and unruly fields involves not only the calculated activity and the strategic establishment of regulatory knowledge and organisational regimes, but also a great deal of experimentation and improvisation. It is yet unclear if the fluidity and messiness that can be observed in contemporary constellations are just a difficult passage to new stabilities and certainties, or – to return to the quote of Zygmunt Bauman at the very beginning of our chapter – if uncertainty, and indeed fluidity and messiness are here to stay.

Now, what are the theoretical and practical consequences that emerge from such conditions of endemic uncertainty? It seems that the dominant Western-style pharmaceutical model of drug development, which was based on the primacy of scientific fact finding and the production of stable scientific evidence, has thus been challenged and indeed destabilised. As many examples throughout this chapter suggest, the question whether stem cell therapies will work in practice has ceased to be centred on scientific evidence only. Rather, various, partially overlapping indications, such as clinical effectiveness, ethical acceptability, commercial viability, market acceptance, and so on, form part of unstable and cobbled innovating practices, rather than pre-established innovation models. We will need to further explore the meanings and the practical consequences of this emerging constellation that seems to challenge the idea that innovation can be rationally promoted and fostered by government agencies.

Moreover, embracing radical endemic uncertainty and complexity will bring into focus the very political dimensions of governing science, technology, and innovation, as decisions need to be taken even if there might not be sufficient scientific evidence, clear legal frameworks, or stable social institutions in which these decisions can be grounded and justified.

Notes

1. Stem cell research has, for various reasons, also become an intensively studied area of research for social scientists of various disciplines. Major

contributions to this matter can be found in Gottweis (2002), Waldby (2002), Jasanoff (2005), Prainsack (2006), Metzler (2007), Geesink et al. (2008), Gottweis et al. (2009).

2. By contrast, a broad range of social studies of risk focus on the multiplication, fragmentation, and individualisation of technologies of risk salient in the neoliberal restructuring of societies (Dean, 2010) or in the context of the regulatory state and the rise of regulatory politics as the predominant mode of governance (Hood, 2002; Rothstein et al., 2006).

3. Drawing on some work from governmentality literature, we understand governing broadly as a purposeful practice of conducting, managing, ordering, regulating, and/or controlling an object or an ensemble of objects. However, 'purposeful' does not say anything about the relation between intent and outcomes. It is in this sense that governing must be seen as a practice, based on some knowledge of the object to be governed, which says nothing about its outcomes or effects (Rose and Miller, 1992; Dean, 2010).

4. There is an ongoing and lively debate over the transformation of the state in the context of globalisation we cannot immerse in at this point. What is common in most perspectives is to understand the state as a mobile and instable set of relations rather than a fixed entity or actor with clear-cut boundaries.

5. It is important to avoid any cynicism here: the assertion that ethics form part of larger political, commercial, and economic strategies does not mean that all bio or research ethics can be reduced to simple 'ideological' manoeuvres involved in the creation of a biomedical hegemony of some stakeholders vis-à-vis others. However, ethics is in itself an embattled terrain hegemonised by particular actors – and not a critical frame of reflection *exterior* to the power struggles that constitute stem cell research and therapy.

6. The irritations and hybridisations caused by tissue engineering have been well studied from scholars of science, technology and society. See, for instance, Brown et al. (2006), Faulkner (2009), and Hogle (2009).

7. The FDA, of course, has also been the place and the target of various serious public conflicts and struggles, accused sometimes for 'overprotecting' the public, and hence slowing down the drug approval process, sometimes for perfunctory control that leads to the marketing of unsafe medicines.

8. Scholars in a science studies tradition might be cautious to accept this portrayal of complexity by studying the social and historical processes by which Aspirin has become such an undercomplex, well-manageable chemical object. Another line of thought might lead to the nuanced question if these different assessments of structural complexity and messiness have something to do with a more global transformation of socio-epistemic and cultural configurations. In these lines, a chemical or biomedical entity would emerge differently in a modernist, structured, and highly disciplined culture and in a 'liquid' post-modern society characterised by fluid forms of control and governance (Deleuze, 1992; Baumann, 2000; Law, 2004).

9. See also Cambrosio et al. (2006) on a general discussion of 'regulatory objectivity' in (bio-)medicine.

10. For a comprehensive analysis of the regime-building process that led to the advanced therapies regulation in Europe, see Faulkner (2009).

11. It is significant to note that the authors explicitly mention both the 'more advanced' adult cell-based and embryonic stem cell-based therapies. It indicates that the sole focus on embryonic stem cells has given way to a more plural universe that has been strategically consolidated as the 'regenerative medicine industry' in the course of the last decade.

12. The unethical side of stem cell treatments is constructed as an ethical, regulatory, and political problem by unstable and informal alliances of researchers, science journalists, and most notably bioethicists. This problematisation is most prominently elaborated in accounts of so-called 'stem cell tourism', covered in high-profile media such as *Nature* and *Science*. For a detailed bioethical discussion see the Special Issue of the *American Journal of Bioethics*, 10 (5), 2010.

13. The university clinic has always been characterised by the constitutive coincidence of experimental research and clinical care, blurring the boundaries of science and health care, physician and researcher, patient and experimental subject. It is in the last few decades that the clinic has become the object of governmental purification processes that seek to draw clear boundaries between research and care. We cannot go into these fascinating observations here, but we need to suggest that in experimental stem cell therapy, these boundaries are again problematised and partially subverted in practice, while there exist tremendous efforts to keep them intact and alive.

14. The principal investigator of the team told us in an interview that the estimated costs of completing the aspired Phase II trial were about 25 million euros. In an article, Eisenstein and colleagues estimate the costs of conventional Phase III trials in cardiology to be much higher: about 83 million US dollars for acute coronary syndromes, and 142 US dollars for congestive heart failure (cf. Eisenstein and Lemons, 2005). One of the major cost factors in clinical trials is the number of patients enrolled in the study.

References

Anon. (1997). Genzyme's carticel gets FDA panel OK. *The pharma letter*, Available at: http://www.thepharmaletter.com/file/57662/genzymes-carticel-gets-fda-panel-ok.html, accessed 16 November 2011.

Baker, M. (2008) Stem cell society urges action on bogus clinics, *Nature*, doi:10.1038/news.2008.1276, 3 December.

Baker, M. (2009) Melissa Carpenter: making stem cells for many, safely, *Nature Reports Stem Cells*, doi:10.1038/stemcells.2009.113.

Barry, A. (2001) *Political Machines: Governing a Technological Society.* London: Athlone.

Bartunek, J. et al. (2006) The consensus of the task force of the European Society of Cardiology concerning the clinical investigation of the use of autologous adult stem cells for the repair of the heart, *European Heart Journal*, 27: 1338–1340.

Bauman, Z. (2000) *Liquid Modernity.* Cambridge: Polity Press.

Bauman, Z. (2007) *Liquid Times: Living in an Age of Uncertainty.* Cambridge: Polity Press.

Beck, U. (1999) *World Risk Society.* Cambridge: Polity Press.

BIONET (2010) *Ethical Governance of Biological and Biomedical Research – Chinese and European Collaboration Project Final Report*. London: London School of Economics.

Brindley, D. and C. Mason (2012). Human embryonic stem cell therapy in the post-Geron era, *Regenerative Medicine* 7(1): 17–18.

Brown, N., A. Faulkner, J. Kent and M. Michael (2006) Regulating hybrids: 'Making a Mess' and 'Cleaning up' in tissue engineering and transpecies transplantation, *Social Theory and Health*, 4(1): 1–24.

Cambrosio, A., P. Keating, T. Schlich and G. Weisz (2006) Regulatory objectivity and the generation and management of evidence in medicine, *Social Science and Medicine*, 63(1): 189–199.

Carnoy, M., and M. Castells (2001) Globalization, the knowledge society, and the network state: poulantzas at the millennium, *Global Networks*, 1(1): 1–18.

Carpenter, D. P. (2010) *Reputation and Power: Organizational Image and Pharmaceutical Regulation at the FDA*. Princeton: Princeton University Press.

Castells, M. (1996) *The Rise of the Network Society, the Information Age: Economy, Society and Culture*, Vol. I. Cambridge and Oxford: Blackwell.

Ceosz, (2011) Premier Wen Jiabao Visits Beike Biotech [in Chinese]. Available at: http://www.ceosz.cn/Special/2011/201106/2011-06-27/Special_20110627161518_84247.html, accessed 13 May 2011.

Chen, H. (2009) Stem cell governance in China: from bench to bedside? *New Genetics and Society*, 28(3): 267–282.

Chen, H. and H. Gottweis (2011) Stem cell treatment in China: rethinking the patient role in the global bio-economy, *Bioethics*, doi:10.1111/j.1467-8519.2011.01929.x.

Cooper, M. (2008) *Life As Surplus: Biotechnology and Capitalism in the Neoliberal Era*. Seattle and London: University of Washington Press.

Cyranoski, D. (2009) Stem-cell therapy faces more scrutiny in China, *Nature*, 459(2009): 146–147.

Daemmrich, A. and J. Radin (2007) Introduction: historical and contemporary perspectives on the FDA, in A. Daemmrich and J. Radin (eds) *Perspectives on Risk and Regulation: The FDA at 100*. Philadelphia: Chemical Heritage Foundation, pp. 3–13.

Dean, M. (2010) *Governmentality: Power and Rule in Modern Society*. London: Sage.

Deleuze, G. (1992) Postscript on the societies of control, *October*, 59: 3–7.

Eisenstein, E. and P. Lemons (2005) Reducing the costs of phase III cardiovascular clinical trials, *American Heart Journal*, 149(3): 482–488.

Elliot, E. and T. Lemmens (2005) Ethics for sale: for-profit ethical review, coming to a clinical trial near you, *Slate Magazine*. Available at: http://www.slate.com/articles/health_and_science/medical_examiner/2005/12/ethics_for_sale.html, accessed 25 March 2012.

Emanuel, E. J., D. Wendler and C. Grady (2000) What makes clinical research ethical? *Journal of the American Medical Association*, 283(20): 2701–2711.

Emanuel, E. J., T. Lemmens and C. Elliot (2006) Should society allow research ethics boards to be run as for-profit enterprises? *PLoS Medicine*, 3(7): e309.

European Medicines Agency (EMA) (2010) Workshop on stem cell based therapies. Available at: http://www.ema.europa.eu/ema/index.jsp?curl=pages/news_and_events/q_and_a/q_and_a_detail_000095.jsp&mid=WC0b01ac0580294392&jsenabled=true, accessed 25 March 2012.

Faulkner, A. (2009) Regulatory policy as innovation: constructing rules of engagement for a technological zone of tissue engineering in the European Union, *Research Policy*, 38: 637–646.

Fink, D. W. (2009) FDA regulation of stem-cell based products, *Science*, 324: 1662–1663.

Fischer, F. (2003) *Reframing Public Policy: Discursive Politics and Deliberative Practices*. Oxford: Oxford University Press.

Fischer, F. and J. Forester (eds) (1993) *The Argumentative Turn in Policy Analysis and Planning*. Durham and London: Duke University Press.

Fischer, F. and H. Gottweis (eds) (2012) *The Argumentative Turn Revisited: Public Policy as Communicative Practice*. Durham and London: Duke University Press.

Food and Drug Administration (FDA) (1997) Reinventing the regulation of human tissue. Available at: http://www.fda.gov/BiologicsBloodVaccines/TissueTissueProducts/RegulationofTissues/ucm136967.htm, accessed 25 March 2012.

Franklin, S. (2003) Ethical biocapital: new strategies of cell culture, in S. Franklin and M. Lock (eds) *Remaking Life and Death: Toward an Anthropology of the Biosciences*. Santa Fe: School of American Research Press, pp. 97–128.

Ganchoff, C. (2008) Speaking for stem cells: biomedical activism and emerging forms of patienthood, in S. M. Chambré and M. Goldner (eds) *Patients, Consumers and Civil Society (Advances in Medical Sociology Vol. 10)*. Bingley: Emerald Group Publishing, pp. 225–245.

Geesink, I., B. Prainsack and S. Franklin (2008) Stem cell stories 1998–2008, *Science as Culture*, 17(1): 1–11.

Gottweis, H. (2002) Stem cell policies in the United States and in Germany: between bioethics and regulation, *Policy Studies Journal*, 30(4): 444–469.

Gottweis, H., B. Salter and C. Waldby (2009) *The Global Politics of Human Embryonic Stem Cell Science*. London: Palgrave Macmillan.

Halme, D. G. and D. A. Kessler (2006) FDA regulation of stem-cell-based therapies, *Nature Medicine*, 355(16): 1730–1735.

Hardt, M. and T. Negri (2000) *Empire*. Cambridge, MA: Harvard University Press.

Hogle, L. F. (2009) Pragmatic objectivity and the standardization of engineered tissues, *Social Studies of Science*, 39(5): 717–742.

Hood, C. (2002) The risk game and the blame game, *Government and Opposition*, 37(1): 15–37.

International Society of Stem Cell Research (ISSCR) (2008) Patient handbook on stem cell therapies. Available at: http://www.isscr.org/clinical_trans/pdfs/ISSCRPatientHandbook.pdf, accessed 25 March 2012.

Jasanoff, S. (2005) *Designs on Nature: Science and Democracy in Europe and the United States*. Princeton and Oxford: Princeton University Press.

Jessop, B. (1998) The rise of governance and the risks of failure: the case of economic development, *International Social Science Journal*, 50(155): 29–45.

Kent, J., A. Faulkner, I. Geesink and D. Fitzpatrick (2006) Culturing cells, reproducing and regulating the self. *Body and Society*, 12(2): 1–23.

Kiatpongsan, S. and D. Sipp (2009) Monitoring and regulating offshore stem cell clinics, *Science*, 323(5921): 1564–1565.

Landzelius, K. (2006) Patients organization movements and new metamorphoses in patienthood, *Social Science and Medicine*, 62: 529–537.

Langstrupp, H. (2011) Interpellating patients as users: patient associations and the projectness of stem cell research. *Science, Technology and Human Values*, 36(4): 573–594.

Lau, D. et al. (2008) Stem cell clinics online: the direct-to-consumer portrayal of stem cell medicine, *Cell Stem Cell*, 3(6): 591–594.

Law, J. (2004) *After Method: Mess in Social Science Research*. London and New York: Routledge.

Lindvall, O. and I. Hyun (2009) Medical innovation versus stem cell tourism, *Science*, 324(5935): 1664–1665.

MacReady, N. (2009) The murky ethics of stem-cell tourism, *The Lancet Oncology*, 10: 317–318.

Malpas, J. and G. Wickham (1995) Governance and failure: on the limits of sociology, *Journal of Sociology*, 31(3): 37–50.

Marchart, O. (2007) *Post-foundational Political Thought: Political Difference in Nancy, Lefort, Badiou and Laclau*. Edinburgh: Edinburgh University Press.

Martin, J. (2005) Collaboration in cardiovascular stem-cell research, *The Lancet*, 365(9477): 2070–2071.

Martin, J. (2009) Is there a future for gene therapy? Keynote address at the London Regenerative Medicine Network meeting, University College London, February.

Mason, C. and E. Manzotti (2010) Defeating stem cell tourism, *Regenerative Medicine*, 5(5): 681–686.

Mendick, R. (2011) Stem cell law loopholes allow XCell-Center to operate in Germany, *The Telegraph Online*, 24 October 2011. Available at: http://www.telegraph.co.uk/health/healthnews/8082925/Stem-cell-law-loopholes-allow-XCell-Center-to-operate-in-Germany.html, accessed 25 March 2012.

Metzler, I. (2007) 'Nationalizing embryos': the politics of human embryonic stem cell research in Italy. *Biosocieties*, 2(4): 413–427.

Mirowski, P. and R. Van Horn (2005) The contract research organization and the commercialization of scientific research, *Social Studies of Science*, 35(4): 503–548.

Morrison, M. (2012) Promissory futures and possible pasts: the dynamics of contemporary expectations in regenerative medicine, *BioSocieties*, 7(1): 3–22.

Office of the Press Secretary (1997) Fact Sheet on FDA. Available at: http://archives.clintonpresidentialcenter.org/?u=112197-fact-sheet-on-fda.htm, accessed 25 March 2012.

Osiris (2012) Therapeutics – Prochymal. Available at: http://www.osiris.com/therapeutics.php, accessed 13 May 2012.

Pollack, A. (2010) Stem cell trial wins approval of F.D.A., *The New York Times*, 30 July 2010. Available at: http://www.nytimes.com/2010/07/31/health/research/31stem.html, accessed 25 March 2012.

Pollack, A. (2011) Geron is shutting down its stem cell clinical trial, *The New York Times*, 14 November 2011. Available at: http://www.nytimes.com/2011/11/15/business/geron-is-shutting-down-its-stem-cell-clinical-trial.html, accessed 25 March 2012.

Prainsack, B. (2006) 'Negotiating life': the regulation of human cloning and embryonic stem cell research in Israel, *Social Studies of Science*, 36(2): 173–205.

Roberts, M. (2010) First trial of embryonic stem cells in humans, *BBC News*, 11 October 2010. Available at: http://www.bbc.co.uk/news/health-11517680, accessed 25 March 2012.

Rose, N. and P. Miller (1992) Political power beyond the state: problematics of government, *British Journal of Sociology*, 43(2): 173–205.

Rothstein, H., M. Huber and G. Gaskell (2006) A theory of risk colonisation: the spiralling regulatory logics of societal and institutional risk, *Economy and Society*, 35(1): 91–112.

Salmikangas, P. (2010) Stem cell-based medicinal products as advanced-therapy medicinal products in the European Union: reflection paper on stem-cell-based medicinal products. Paper presented at the EMA workshop on stem cell clinical trials, European Medicines Agency, London, 10 May.

Salter, B. and S. Salter (2007) Bioethics and the global moral economy: the cultural politics of human embryonic stem cell science, *Science, Technology, and Human Values*, 32(5): 554–581.

Schächinger, V. et al. (2006) Intracoronary bone marrow derived progenitor cells in acute myocardial infarction, *New England Journal of Medicine*, 355(12): 1210–1221.

Strauer, B. et al. (2002) Repair of infarcted myocardium by autologous intracoronary mononuclear bone marrow transplantation in humans. *Circulation*, 106: 1913–1918.

Sunder Rajan, K. (2006) *Biocapital: The Constitution of Postgenomic Life*. Durham and London: Duke University Press.

Thacker, E. (2005) *The Global Genome: Biotechnology, Politics, and Culture*. Cambridge, MA: MIT Press.

Vermeulen, N., S. Tamminen and A. Webster (eds) (2012) *Bio-objects: Life in the 21st Century*. London: Ashgate.

Wahlberg, A. and T. Streitfellner (2009) Tourisme de cellules souches, désespoir et pouvoir des nouvelles thérapies [Stem cell tourism, desperation and the governing of new therapies], in A. Leibing and V. Tournay (eds) *Technologies de l'espoir. les débats publics autour de l'innovation médicale – un objet anthropologique a définir*. Québec: Presses Universitaires de Laval, pp. 645–66.

Waldby, C. (2002) Stem cells, tissue cultures, and the production of biovalue, *Health*, 6(3): 305–323.

Webster, A. (2013) Introduction, in A. Webster (ed) *The Global Dynamics of Regenerative Medicine: A Social Science Critique*. London: Palgrave Macmillan, pp. 1–17.

Webster, A., C. Haddad and C. Waldby (2011) Experimental heterogeneity and standardization: stem cell products and the clinical trials process, *BioSocieties*, 6(4): 401–419.

Wilson-Kovacs, D. M., S. Weber and C. Hauskeller (2010) Stem cells clinical trials for cardiac repair: regulation as practical accomplishment, *Sociology of Health and Illness*, 32(1): 89–105.

Wittlesey, K., M. Lee, J. Dang and M. Colehour (2011) Overview of the FDA regulatory process, in A. Atala, R. Lanza, J. A. Thomson and R. M. Nerem (eds) *Principles of Regenerative Medicine*, 2nd ed. London: Elsevier Academic Press.

Zeng, K. Y. (2009) Stem cell therapy today in the People's Republic: an interview with Sean Hu, of Beike Biotechnology, *h+ Magazine*, 28 April.

5
Procuring Tissue: Regenerative Medicine, Oocyte Mobilisation, and Feminist Politics

Susanne Schultz and Kathrin Braun

Introduction

Women's bodies have come to occupy a critical position at the intersection of bioethical debates, regulatory politics, and biomedical research strategies of regenerative medicine in recent years since they are capable of generating a variety of tissues which are of particular interest for a type of medicine arising from the novel regenerative paradigm. Regenerative medicine, as explained in the introduction, aspires to incite, deploy, and control the body's own capacity for self-repair rather than merely acting on it through applying drugs, medical devices, organs, or tissue. And it does so, or seeks to do so, via harnessing the regenerative potential of certain types of tissues the generation of which critically involves women's bodies, such as oocytes, embryos, aborted foetuses, or umbilical cord blood. Stem cell lines have been derived from human embryos, aborted foetuses, and umbilical cord blood already, but up to the emergence of induced pluripotent stem cells (iPS cells), it looked like human oocytes were irreplaceable for stem cell research strategies that aspired at generating patient-specific cells, tissue or, in the long run, organs.

Up to now, there are – theoretically – three such strategies: iPS cells, parthenogenic stem cells, and somatic cell nuclear transfer (SCNT). iPS cells go back to a procedure introduced by Shinya Yamanaka in 2006 through which somatic cells, for instance skin or fat cells, are reprogrammed into being pluripotent and thus similar to embryonic stem cells. They are regarded by many as being ethically less problematic than human embryonic stem cells or SCNT, because no oocytes or embryos are required for their generation. To date, however, it is still not clear exactly how far iPS cells resemble natural stem cells and how great the risk is that they cause tumours. Parthenogenic stem cells are embryonic-like stem cells derived from an unfertilised oocyte, which has been activated without sperm and induced to divide as if it had been fertilised. In January 2009, the International Stem Cell Corporation (ISCO), a Californian biotech firm, announced they had created layers of retinal progenitor cells from human parthenogenic stem cells and had transplanted it into animals for testing.[1] In SCNT, by contrast, also known as research cloning or therapeutic cloning, the nucleus of a somatic cell is transferred into a denucleated oocyte, which is then induced to develop into an embryonic stage of a blastocyst. Ideally, researchers would then derive stem cells and genetically customised cells, tissue, or even organs from that blastocyst that would not be rejected by the recipient's body. In 2008, the Californian biotech firm Stemagen announced they had managed to create the first cloned human blastocyst.[2] Already in 2005, however, a team at Newcastle University in the United Kingdom claimed they had created a human blastocyst through nuclear transfer, using failed-to-fertilise eggs from in vitro fertilisation (IVF) (Stojkovic et al., 2005). In either case, however, the goal of establishing embryonic stem cell lines from cloned embryos has not been accomplished up to now (French et al., 2008).

Nevertheless, recent developments in both research and regulation indicate that interest in human oocytes and their regenerative potential is still strong: in October 2011, a team led by Dieter Egli at Scott Noggle of the New York Stem Cell Foundation reported they had – by accident – found out that when the nucleus of the oocyte was *not* removed, as in SCNT, the resulting triploid cells developed more easily to the blastocyst stage (Noggle et al., 2011); 63 oocytes were needed here to establish one normal set of cells (Wade, 2011). While these cells are far from having therapeutic value at the moment, it is not unlikely that the report will reinvigorate research requiring human oocytes. Also, it is arguably no coincidence that the experiment was performed in the state of New York, where researchers are permitted to offer women up to $10,000 public money per menstrual cycle for their oocytes,[3] a transaction scheme we will return to. For now it looks like human oocytes will likely remain what they currently are: one of the most 'touchy' and contested bio-objects in regenerative medicine. They are touchy objects in that – unlike many other bio-objects – they cannot be divided or multiplied nor easily be preserved, stored, and shipped all over the world. Therefore, for anyone dealing with oocytes, negotiating long distances in terms of time and space is an issue. In order to access and process oocytes, stem cell researchers need close temporal and spatial proximity to the IVF sector, where the overwhelming part of oocyte retrieval takes place, *or* they need the means to motivate women to undergo egg retrieval solely for research purposes – which works best by offering material incentives. Oocytes are touchy too inasmuch as they are ethically, legally, and politically extremely contested, especially in regard to the procurement for research. They have become the subject of a global controversy in recent years, which can be read as a textbook debate over the new bioeconomic appropriation of body materials and its social and ethical implications.

The debate was triggered not least by the so-called Hwang scandal a few years ago (Leem and Park, 2008; Tsuge and Hong, 2011). According to South Korea's National Bioethics Committee, the South Korean stem cell researcher Woo Suk Hwang had obtained a total of 2221 oocytes from 121 women for his experiments, published in *Science*, which aimed at extracting stem cells from cloned human embryos. Of those women, 85 had been paid for contributing their oocytes and 2 of those who had not received payment had been junior researchers at the time on Hwang's team and can thus be considered as having been in a dependent position (Leem, 2008).[4] The story fuelled concern that SCNT research might significantly increase the need for oocytes and thus scale up the appropriation of women's body materials and establish a new form of women's exploitation. Feminists were especially engaged in this debate, as civil society actors, experts, and feminist theorists.[5] A major issue at stake here was the question of how the exploitation of women on a global scale might be prevented – especially in the light of the health risks involved in egg harvesting.[6]

This chapter presents a critical review of feminist perspectives and positions concerning oocytes for research. Although oocyte procurement for research purposes arguably is a highly specialised practice and as yet still relatively rare, at least in Europe (Braun and Schultz, 2012), it is also an issue that has triggered fundamental considerations and controversies concerning the conceptualisation of bodily integrity in the age of biotechnology in general and the bioeconomy of the female body, as well as women's bodily self-determination in particular. Thus, feminist debate on this issue is at the same time a very particular yet also a very fundamental and general one which merits a systematic review. We will argue that we can discern three main political approaches, each based on a specific way of framing women's bodily existence and the relationship between the woman, her body, and bodily materials and, correspondingly women's subject position within the bioeconomy. We will argue that in general the debate tends to remain confined to a specific discursive framework focusing on the moral status of the individual body on the one hand and the bio-object (the oocyte) on the other, and the resulting rights of the oocyte provider in relation to her body and her oocytes. We argue that this focus at the same time individualises and universalises the issue of oocyte procurement, approaching it mainly as a matter of conceptualising and assessing the moral status of bodily process and biological 'objects' – in this case the oocyte – and basing political positions on these conceptions. We propose that feminist analyses move beyond this framework and shift the focus from the individual woman, her body, and her oocytes to processes and practices of 'doing bodies' in their specific contexts and circumstances. Such shift of focus would imply in our view that we redirect attention to social relationships constituted, formed, and transformed by practices of oocyte procurement and the specific bioeconomic rationalities in which they are embedded. We thus seek to strengthen a perspective of social and anti-capitalist critique within feminist debates on the new bioeconomy, one that scrutinises and challenges the tendencies towards appropriating the female body rather than seeking to regulate them. In order to do so, we refer to the strand of anti-racist, anti-capitalist, and anti-eugenic women's health movements that have formed in opposition against global biopolitical population policy, particularly their conceptualisation of reproductive rights. We draw from this lineage in order to develop a contextualising position regarding women's bodily involvement within the new bioeconomy.

Before entering the discussion among feminists on the issue, however, we would like to point out certain regulatory and economic developments in recent years that, in our view, form part of the socio-economic and political context within which the issue should be considered.

Trends towards commercialisation

On the one hand, it may seem that the controversy about oocytes for stem cell research has lost momentum in recent years, which may reflect the fact that concerns about exploding demands and the emergence of a global market for oocytes have not materialised so far. The hype about research cloning had not least been fuelled by Hwang and his announcement of an alleged breakthrough, which has now turned out to be a fraud: the purported stem cell lines allegedly created through cloning never existed (Gottweis and Triendl, 2006). Until now, no further research team has announced the successful extraction of stem cells from cloned human embryos. There has been some progress in developing cloning techniques for research but no big breakthrough. As of early 2010 there were only three research sites in Europe where research into human SCNT took place (Braun and Schultz, 2012). Instead, stem cell researchers have invested considerable effort in research on iPS cells as a means of creating patient-specific and immuno-compatible tissue that does not rely on human embryos or oocytes.

On the other hand, even though research cloning currently plays only a minor role in stem cell research, this does not mean that the issue of oocyte procurement and its implications for women as oocyte providers is obsolete. Bioeconomic investments in research may be reallocated at any time, for instance, if iPS cells turn out not to work or if efforts to establish human stem cell lines through SCNT prove successful after all. Moreover, recent events in policy and regulation suggest that there are still forces pushing for a commercialisation of oocytes for research, either openly or behind the scenes, thus preparing the ground for research cloning through restructuring the bioeconomic terrain within which it is embedded. For instance, as noted above, in June 2009, the Empire State Stem Cell Board of the State of New York permitted stem cell research to use public research funding to pay women as much as $10,000 per menstrual cycle for giving their eggs to research (NYSTEM, 2009; NYT, 2009).

In Europe, the Human Fertilisation and Embryology Authority (HFEA) in the United Kingdom also made a move towards commercialising oocytes. In October 2011, the authority decided to change its policy of compensating sperm and egg 'donors'. Hitherto, women who underwent oocyte retrieval and gave the eggs either for reproductive or for research purposes could only claim reimbursement for out of pocket expenses and loss of earning of up to £250. The new scheme will allow clinics to hand egg donors a fixed sum of £750 per cycle of donation (HFEA, 2011). While this is still called a form of 'compensation', and not 'payment', in contrast to the former reimbursement scheme it means that women now can make a real, if moderate, financial gain by undergoing egg retrieval. Thereby, the United Kingdom has adopted a scheme which has been common in Spain already, where women can gain €1000 or even more per cycle in exchange for their eggs (Dickenson and Alkorta Idiakez, 2008; Alkorta Idiakez, 2010; Braun and Schultz, 2012).

Even in countries where payment for oocyte 'donation' is prohibited by law, there is a trend towards more or less indirect forms of transactions (Braun and Schultz, 2012). Various schemes of transactions are currently in use. One of them, which we termed the reallocation model, has been employed by an SCNT research project conducted by Anna Veiga at the Centro de Medicina Regenerativa de Barcelona. The project uses oocytes from women recruited by a cooperating private fertility clinic. These women received €900 per 'harvest' in exchange of undergoing hormone stimulation and oocyte retrieval. Originally, the eggs had been given for IVF purposes, but Veiga secured approval from the review board to reallocate some eggs for research.[7] In the flourishing private IVF industry in Spain (Pavone and Arias, 2012), this amount is considered adequate 'compensación' for supplying oocytes. Mostly it is immigrants from Eastern Europe and Latin America, students and workers from the poorer social strata who are willing to undergo these strenuous procedures.

In the United Kingdom, Alison Murdoch, researcher at the North East England Stem Cell Institute (NESCI) in Newcastle upon Tyne, uses oocytes from IVF patients who then receive a discount on their own IVF treatment in return. This type of scheme is usually known as egg sharing.[8] Meanwhile, the HFEA has declared that not only a reduction in IVF treatment costs but also other benefits in kind such as storage services and moving up the waiting list may be offered in exchange for oocytes (HFEA, 2011). We find similar strategies of using financial incentives in California, which has also banned commercial oocyte trade for research purposes (Schultz, 2010). In today's complex tissue economy, as Waldby and Mitchell (2006) argue, the boundaries between a gift economy with purely altruistic donations of body materials on the one hand and a market economy where body materials are sold in exchange for money are often blurred. Gift and commodity systems are not mutually exclusive any more, if they ever were, but intertwined and overlapping (Waldby and Mitchell, 2006). Regulatory systems governing oocyte procurement tend to form a case in point here, accommodating a range of hybrid transaction schemes that are neither purely altruistic nor purely commercial – while at the same time formally upholding a general principle of non-commercialisation. Oocyte procurement for research, in Europe, grows out of a socio-ethico-political environment that combines such favourable regulatory frameworks as outlined above with close institutional, personal, and logistical connections between stem cell research on the one hand and a well-established – mostly private – fertility industry on the other (see Braun and Schultz, 2012; Pavone and Arias, 2012).

In the following section, we will review how feminists scholars have discussed oocytes for research in the past years, particularly concerning issues of health risks for oocyte providers, potential exploitation, and commercialisation of the female body. Within this debate, we think we can discern three main paradigms regarding the way women's subject position is constructed in relation to oocyte procurement in particular and the position of the female body in the new bioeconomy in general. We suggest that the female subject in this debate figures mainly as the body-owner, the regenerative labourer, or the potential victim, with each figure being tied up with certain assumptions about the nature of the problem at stake as well as of the required course of action in order to tackle it. Note, however, that the actual lines of reasoning in feminist contributions do not necessarily follow only one of these subject constructions; the picture may combine features of more than one figure and these may partly overlap or intermingle. Still, we would hold that sorting out the main features of the dominant subject constructions may help to structure the debate and take it further.

Owner, rights-bearer, and the contractual moment

One line of reasoning sketches the oocyte provider primarily as the autonomous agent who acts as a contracting party within different arrangements of oocyte transactions. Reflecting classical liberal notions of the individual and his or her relation to others as well as to his or her body and bodily existence, this conception forms the basis for much of present-day regulation policy and regulatory demands in the issue area of biomedicine and bioeconomics. In essence, these policies, taking the shape of bioethical or medical guidelines, domestic law, or international agreements, aim at guaranteeing individual rights and freedoms such as individual self-determination and freedom of choice. Oocyte provision is conceived as a contractual relationship which in principle is a matter of free decision-making by autonomous agents and the major challenge therefore is to ensure that individual decision-making is indeed free and undistorted and based on sufficient information, the main mechanism being informed consent. While there is much debate on and disagreement about what exactly informed consent should or should not include, the common characteristics of this approach are its assumption of individuals as free and autonomous subjects, its emphasis on rights and obligations, and its focus on the contractual moment. For a liberal approach, concluding the contract is the critical moment and at the heart of the contractual moment is the act of giving informed consent. Feminists arguing along these lines insist on women's agency and their capacity to make free and responsible decisions on their own behalf on the one hand and on institutional mechanisms to protect the individual rights and interests of women on the other, with the main focus being on informed consent and the contractual moment. Advocating such a feminist-liberal approach, Bonnie Steinbock (2209), for instance, defends the New York State decision to allow paying women for undergoing oocyte retrieval solely for research purposes. While undue inducement certainly is an issue to be considered, she argues, the risks involved in research 'donation' do not differ from those in reproductive 'donation'. Since paying women for reproductive egg 'donation' has become common, and legally permitted, in the United States for a long time, it would be unjustified to treat research 'donation' differently.[9] 'Moreover', she argues, 'women who are given full information about the risks and burdens are surely capable of making the decision to donate for themselves.'[10] Banning women from selling their eggs from this perspective is deemed a case of overprotection veering on paternalism.

The main concerns, then, are revolving around the question of what information women require for making a truly autonomous decision. What information must be given at what point in time, by whom, under which circumstance, how detailed, comprehensive, and comprehensible does it have to be? There is consensus in the literature that women must be informed about the course of the procedure and the burdens and health risks involved, although there may be differences as to the how detailed this information actually has to be.[11]

In contrast, there is much less debate about what happens with the eggs *after* they are extracted: Who holds the property rights during the research process? Who makes the decisions regarding the commercial use of the research results? And who is entitled to share in the profits? Standard informed consent procedures assume that signing the official form amounts to fully surrendering property rights to the research entity. Donna Dickenson has elaborated on this neglected dimension, pointing at possible kinds of protection informed consent should offer *after* the point of oocyte retrieval. 'Consent is normally conceived as consent to the initial procedure, not to "downstream" uses of the tissue: as a one-off requirement rather than an ongoing set of powers and duties' (Dickenson, 2007, p. 18). Questions such as who holds the property rights during the research process, who decides about the commercial use of research results, are rather bracketed out. Referring to a Lockean concept of property in the body, Dickenson suggests an interpretation of the liberal contract in which property flows from the individual's property in her own person rather than from property in her body.[12] Starting from the concept of self-ownership, Dickenson argues, we do not need to assume that I own my physical body like a thing. Since selfhood is always embodied selfhood, rights in relation to my body thus flow from embodied self-ownership. Thus, women's rights in relation to their oocytes in Dickenson have a twofold foundation: as long as they have not been separated from my physical body they are covered by my embodied self-ownership. The process of separating them from my body, on the other hand, certainly involves my labour and action. It mixes my body materials with my labour, to speak with Locke, thus making them the product of my labour and constituting another source of my property rights towards them. Like Waldby (see below), Dickenson conceptualises the oocyte as a product of women's labour (Dickenson, 2006). Aside from this foundational reasoning, she usefully suggests the need to think of property not as a single coherent concept that does or does not apply but as a bundle of different rights that can be disaggregated from each other. Concerning the question of property in the body she suggests that we 'think long and hard about which rights we want to protect' (Dickenson, 2007, p. 12). The bundle of rights and obligations may include rights of using, managing, or transmitting body materials as well as post-transaction rights such as a right to income derived from its use by others. Based on this conception, we can conceive of women selling their oocytes *and* retaining certain rights even upon concluding the contract, for example, control over what happens with their body tissue after retrieval.

Apart from specifying the terms of informed consent, a major concern for liberal approaches to oocytes for research is the question of whether and how financial transactions affect freedom of choice and whether they can function as a means to protect women's rights or interests or, on the contrary, as mechanisms that endanger them. There are different views on this question; some see payment as an effective means to curb the risk of trafficking, exploitation, stratification, and black markets emerging, others view altruistic donation as the best way to prevent these things from happening. Between both positions there is, however, a range of arguments for various modes of reimbursement or compensation. Isasi and Knoppers (2007), for example, introduce various models of financial transactions and opt for a mixed model which they see as a more equitable solution 'wherein there is a statutorily determined amount offered. This set amount would, however, distinguish between the types of reproductive material' (Isasi and Knoppers, 2007, p. 42). Another position is to opt for the reimbursement of direct out-of-pocket expenses (Reynolds and Darnowsky, 2006, p. 19).

Advocating payment, Thompson (2007) argues that women should in fact be paid for providing oocytes and additionally have the right to share in potential profits derived from the respective stem cell research. Payment, in her view, should be designed as a salary for a service provided and as compensation for adverse health effects and risks involved rather than a monetary equivalent of the eggs as such (Thompson, 2007). While Thompson does not deny the problem of undue incitement, she argues that without removing the research as such, prohibiting payment and thereby restricting the availability of oocytes will push rather than curb the development of a black market.

This point however, we would argue, starts from the assumption that research using human oocytes is and will be continuing anyway, whether remuneration for oocyte provision exists or not. Seen against the background of our own research, we think this is a momentous misconception: research strategies requiring oocytes *cannot* be viewed separately from the question of how oocyte procurement is regulated and organised and whether or not mechanisms of remuneration or 'compensation' are available or not. Without monetary incentives, we would argue, this type of research is rather unlikely to persist in the long run since it is extremely difficult for researchers to obtain sufficient oocytes from altruistic donors only (Braun and Schultz, 2012). Mobilising oocytes through some form of payment or other therefore forms a prerequisite for this branch of research – rather than a means to counter adverse effects of a practice that would continue anyway. As regards the problem of undue incitement, Thompson suggests it could be solved by national and international regulation and oversight, ensuring compliance with standards of informed consent. Potential stratifying effects of payment, meaning that mostly poor women would be incited to take serious health risks because they need the money, could be avoided by capping compensation 'at a level that could be earned by other kinds of physically demanding service work' (Thompson, 2007, p. 209).

Other feminists argue that economic incentives inevitably distort free and autonomous decision-making; the commercialisation of body materials should be prohibited in order to eliminate economic incentives. If oocyte donation is permitted at all, it should be altruistic donation only (Schneider, 2006; Baylis and Mcleod, 2007). Schneider points to consequentialist arguments against payment: 'Paying women to donate egg cells may undermine a proper informed consent. The financial incentive may rule out the adequate consideration of medical and psycho-social risks.' And, 'payment can also tempt concealment of important information by the donor, e.g. about infections and diseases, in order not to be rejected as a donor' (Schneider, 2006, p. 232).

Concerning the question of oocyte provision and women's exploitation, Haimes et al. (Haimes et al., 2012) suggest it is important to approach this question from the point of view of 'donors' themselves, thereby framing exploitation as such as being contingent on the question in which way the women involved interpret the contractual situation. The authors investigated the views and experiences of women who took part in the 'Newcastle egg sharing for research' scheme[13] and found that although the women did not deny the possibility of exploitation in the context at stake, they dismissed it in their individual case, emphasising that they had volunteered for the scheme, that they had been well informed, and that they had benefited from it. Approaching the question of exploitation from this angle of course means to reduce it to the question of perceived individual self-determination at the point of the contractual moment.

Overall, the subject position of women, in this strand of contributions, is that of a bearer of rights, the owner of her body and/or her self, equipped with freedom, agency, and the capacity to make decisions in her best interest and to enter contract arrangements concerning her body and body materials. Whether or not she receives payment, she is constructed primarily as a contracting party and the main concern is about securing the undistortedness of the contractual moment. The role assigned to the state or international bodies is to make sure that she is provided adequate information and freedom of choice so that the conditions for concluding the contract are met.

Women as potential victims and powerless research resources

Another strand of feminist activists and scholars, in contrast, questions the contractual paradigm, along with the primacy of informed consent. Feminists following this line of reasoning emphasise the health risks involved in oocyte retrieval and the existence of asymmetric power relations between research interests and women as oocyte providers. Informed consent, here, is deemed a rather problematic and insufficient concept for various reasons. For one thing, as Schneider (2006), for instance, argues, retrieving oocytes for research purposes collides with the fundamental medical ethics principle of *primum non nocere*, especially in the case of women who do not undergo hormone stimulation and oocyte extraction in the course of their own IVF treatment, but solely for research purposes. This practice, as Ingrid Schneider points out, is incompatible with the principle that doctors may not harm their patients or put them in danger in order to benefit a third party (Schneider, 2006). As George (2008 shows, following the reasoning of Magnus and Cho (2005), the analogy to clinical trials or organ donation, which is often made, is misleading. In clinical research, the risks lie in the research itself, not in the process of obtaining indispensable research materials. Further, it is worth noting that clinical research is subject to much more extensive regulation than oocyte extraction, especially in the private IVF sector. In case of living organ donation, the general rule is that the benefit for the recipient has to be considerable whereas the risk for the donor must be very low. In the case of basic research that has yet to prove its usefulness, as is the case with SCNT, the cost–benefit ratio can be assumed to be much less favourable.

Another major argument brought forward against the contractual paradigm is that in face of persisting uncertainty about the potential adverse effects of hormone stimulation, free and informed choice is no meaningful concept whatsoever. Feminists such as Diane Beeson and Abby Lippman (Norsigian, 2005; Beeson and Lippman, 2006; Dickenson and Alkorta Idiakez, 2008; George, 2008; Alkorta Idiakez, 2010) contend that, due to insufficient research, we still do not know enough about the long-term health risks of hormone stimulation for women and their offspring while there is also disagreement among scientists as to the incidence of short-term risks, such as ovarian hyperstimulation syndrome.[14] For these reasons, some feminists have called for a global moratorium on oocytes for research. Unless all the health risks involved in oocyte extraction, hormone stimulation, and surgical intervention, both for women themselves and for their offspring, have not been researched conclusively and found to be acceptable, such research should be stalled they argue (Beeson and Lippman, 2006; Hands off our Ovaries, 2006; Dickenson and Alkorta Idiakez, 2008).

A further line of reasoning concerns the usefulness of research using human oocytes as such. Feminist critics of this research stress that no therapies derived from SCNT have materialised as yet. This, they argue, is important to note since research has a vested interest to create the impression that revolutionary treatments are imminent. Representations of this research, so the critics go, are dedicated to capture public imagination, raise hope and expectation, and thereby mobilise material and immaterial support from governments, funding agencies, ethics committees, potential oocyte providers, and the public in general. The language used, critics argues, is often manipulative and biased, which adds to rendering free and autonomous decision-making illusionary (Beeson and Lippman, 2006; Sexton, 2006). Tsuge and Hong, for instance, show that many women who donated oocytes to Hwang and his research believed they were directly helping to cure seriously ill people (Tsuge and Hong, 2011).

Another key concern from this perspective is that neither payment nor a non-commercialisation policy will successfully rule out that women are exploited. Under conditions of social inequality, be it local or global, even a small compensation can provide a financial incentive to accept health risks. Baylis and McLeod lucidly argue that it will not work since undue inducement does not depend on the amount of money offered but on the situation of the women: In order to avoid undue incitement by means of capping 'one would have to insist that women from poorer nations be paid considerably less for their eggs that women from wealthier nations (or insist that all women be paid the same low amount, an unlikely scenario and an ethically troublesome one (...)' (Baylis and Mcleod, 2007, p. 729).

Altruistic donation, on the other hand, is questionable as well from a feminist standpoint, since it excludes women as oocyte providers from sharing in the potential profits that may be generated with the help of their contribution, while other parties such as researchers, clinics, or the pharmaceutical industry are not required to operate on a purely altruistic basis. Moreover, altruistic donation may be embedded in power relations. Women may face moral pressure from relatives to donate eggs, or feel pressured to do so by public and/or governmental enthusiasm for stem cell research (Beeson and Lippman, 2006; Sexton, 2006; George, 2008). In the case of South Korea, Tsuge and Hong, who explored the motives of South Korean women to donate oocytes for Hwang and his team, found that many donors had been driven by the wish to 'support treatment for sick children' and to express their 'love for Korea' (Tsuge and Hong, 2011, p. 243). Some donors had been under the impression that by donating eggs they could help family members who suffered from disease or disability. Others saw it as a way to make a sacrifice to the nation[15] equivalent to that by young man joining the military. Hence, altruistic donation has to be placed in a context of gendered notions of civic duty, family obligations, and spiritual virtue.

While feminists leaning towards the liberal construction of women as contractual parties emphasise women's agency and freedom of choice, these critics highlight the vulnerability of women especially in relation to powerful research interests. Women are cast as potential victims, their bodies as a resource to be accessed and exploited for research. Because of the power and knowledge differential involved, women are vulnerable and exposed to manipulation, exploitation, and dangers to their health. For Erica Haimes and her co-authors, this picture largely leaves out women's agency: 'There has been a failure, in some accounts of egg provision, to see women's ability to evaluate the gains and losses for themselves, within a clear awareness of wider structural constraints and opportunities, and then to act accordingly, in their own best interests' (Haimes et al., 2012). A similar argument has been made by Nahmann who talked to oocyte providers in Romania, although for reproductive purposes (Nahmann, 2008). Like Haimes and her co-authors, she portraits oocyte providers as economic agents for whom oocyte provision is a means to realise their wishes or plans, whether these are about a baby, new furniture, or paying off their debts. They may have limited options but they seek to make good use of them under the circumstances given.

Aside from the question of agency versus vulnerability, we see another issue here. While we would agree that power relations, interests, manipulation, and knowledge differentials need to be addressed, we would hold that the role assigned to the state and international bodies requires a more cautious problematisation. There is a downside to state protection, and that is surveillance and control. Calling for the state to protect women effectively against exposure to health risks means calling for a system of oversight and control, including apparatuses to collect, store, and process sensitive data about women's health and lives. In order, for instance, for the state to limit the number of stimulated cycles in an individual woman, as, for instance, Schneider (2006) recommends, the state would have to know exactly when, where, and how often a woman has had hormonal stimulation. If, in addition, the state was to prevent the accumulation of risks, this would require a system of risk profiling based on data about other hormonal treatments, medication, drug abuse, and other potential 'risk factors'. Plus, the system would have to extend across borders.

The same applies in principle to the idea of comprehensive longitudinal studies. The rationale for demanding a moratorium on oocytes for research is not least based on the reasoning that we need comprehensive longitudinal studies first before we assess what the risks actually are, and without reliable knowledge about these risks, women can provide no meaningful informed consent. Such research of all possible risks involved, both to women and to their children, would require comprehensive data collection about these women over long periods of time, across borders, and covering anything that might indicate adverse effects such as miscarriages, infertility, or cancer. Any attempt to establish complete scientific certainty would require comprehensive registration and monitoring practices – a problem that has been strangely neglected in the literature so far.

The regenerative labourer and her productive body

Still another feminist approach is proposed by Catherine Waldby and Melinda Cooper who explicitly seek to counter the portrayal of women mainly as victims, as implicit in some of oocytes for research. Instead, they suggest we conceive of the oocyte provider as a regenerative worker[16] and emphasise her agency and productivity (Waldby, 2008; Waldby and Cooper, 2010). Their intention is 'to relocate (. . .) feminized productivity within the circuits of economic value' (Waldby and Cooper, 2010, p. 10), highlighting the productive role of women who provide tissue for biotechnological research and endowing them with rights and negotiating power. In this view, providing oocytes is primarily a form of productive labour. It may take the form of a contractual relation, constituted through informed consent, and it may involve health risks and burdens but, most importantly, it is productive labour. Waldby and Cooper, not unlike Dickenson discussed above, emphasise the indispensable and productive, albeit mostly obscured, part women play in the new bioeconomy.

Waldby insists that egg harvesting involves collaboration on the part of the woman; her body makes an active, crucial, yet unrecognised contribution to stem cell research. This active contribution, so Waldby argues, can and should form the basis for an industrial relations-type of rights, fought for by oocyte providers themselves. Thus, the active, productive contribution of women as regenerative labourers merits recognition, for instance, along the lines of industrial safety rights such as health insurance, preliminary and follow-up care, and safety provisions but also payment.[17] Collective self-organisation of sex workers or persons who serve as subjects for clinical trials in the Global South might provide a model for this type of struggle.[18] Waldby argues that prohibiting remuneration runs the risk of criminalising oocyte procurement and pushing such activities into unregulated black markets (Waldby, 2008). As noted above, we object that demand for oocytes for research is no independent variable but hugely dependent on regulatory frameworks of oocyte procurement. Legalising payment, in our view, is a factor that actively contributes to the creation of demand here.

Importantly, Waldby and Cooper historicise the concept of regenerative labour which they understand as a phenomenon of the post-Fordist era. Much of feminist theory that draws from a materialist tradition of social critique, they convincingly argue, seeks to interpret the relation between the female body and the contemporary bioeconomy within a framework that belongs to the era of Fordism. Likewise, the normative model of non-commercialised tissue procurement and altruistic donation originates in the context of Fordism. The reproductive labour debate in feminism revolved around the scandal that women were confined to the domestic sphere which was constructed as a private sphere of gift relations where women's work did not count as work and remained invisible, unrecognised, and unpaid. Both gift economy and market economy in the Fordist era were regulated and linked to each other by the nation state. Theorising women's contribution to the contemporary bioeconomy as a new form of unpaid reproductive labour, Waldby argues, however, employs the terms of a regulatory regime that no longer exists. To date, the post-Fordist economy is no longer confined to national borders and 'gift and market systems... are losing their distinctiveness' (Waldby and Cooper, 2010, p. 12).

While historicising regenerative labour on the one hand, Waldby and Cooper at the same time build on a more decontextualised notion of regenerative labour that takes the productivity of the female body and the oocyte as a point of departure. Approaching the issue from this angle, however, means to draw from a version of materialism that owes more to naturalism and a materialist, early Marxist version of subject philosophy than to late Marxist concepts of social relations and relations of production as historically specific forms of organising social life – and ultimately as social practices. Starting from the productivity of the body and/or body materials, we would hold, has some problematic, or at least unsatisfactory, naturalising, and universalising implications. A form of naturalism is involved when Waldby and Cooper argue that the reason why the oocyte provider's productivity remains unrecognised lies in the nature of bodily processes:

[I]t is a form of labour not amenable to quantification in linear, abstract units of time and codified tasks; rather it takes place through the complex time of reproductive metabolism, endocrine circulation, and the unfolding of ontogenic processes, recalibrated through assisted reproductive technologies and stem cell technologies.

(Waldby and Cooper, 2010, p. 9)

The argument seems to be that it is the nature of those bodily processes that escapes mechanisms of quantification, measurement in time, calculability, and therefore economic recognition. Inasmuch as this is held to be a distinctive feature of female regenerative labour, it is implicitly contrasted to other, 'conventional' forms of labour to which these features do not apply. Such a distinction, however, not only tends to reiterate the nature – culture dichotomy, placing women once again on the side of nature and seeking the reason for why they are made invisible and disrespected in natural processes. It also implicitly assumes that non-regenerative, conventional, 'male' forms of labour are by nature more amenable to quantification in abstract time. This, however, obscures the fact that it took centuries of discipline and violence to transform concrete labour into abstract labour, with the labour process being fragmented, quantified, and broken down into units of abstract time. We should also be careful to suggest that 'conventional' labour can be separated from the body more easily whereas regenerative labour cannot; this assumption would play down the fact that it is *always* the embodied self who is engaged in any sort of labour; the body is always involved, both in terms of productivity and in terms of bearing the burden and suffering stress.

Moreover, a decontexualising and universalising element comes in concerning the notion of productivity here. Waldby and Cooper's conception of the regenerative worker's productivity conflates different dimensions of oocyte provision: the active involvement of women in medical procedures; the aspect of being bodily exposed to risks and burdens; the distinctive productivity of the female body that generates (though induced by hormone stimulation) mature eggs; and finally the distinctive productivity of the oocyte after extraction, which is utilised by stem cell research. This creates an odd continuum of 'regenerative productivity' ranging from cellular activity to women's activity such as, for instance, organising her life around hormone injections and visits to the clinic, all encompassed by the concept of regenerative labour.

Related to the concept of productivity that binds together cellular processes and social practices is an element of subject philosophy that implies a certain universalising tendency, in particular when taken as the point of departure for political debate. Waldby and Cooper's contribution displays strong similarities to Michael Hardt and Antonio Negri's notion of biological productivity (Hardt and Negri, 2002). Similar to Hardt and Negri's concept of biopolitical productivity, they alternate between two lines of argument claiming new forms of women's exploitation on the one hand and new forms of women's empowerment as female producers on the other, due to new configurations in biosciences and the bioeconomy. Hardt/Negri and Waldby/Cooper converge in attributing the distinctive biopolitical, respectively, regenerative productivity of female labourers to a global post-Fordist social and economic regime. Yet, insisting on an overarching concept of productivity that would cover both social and biological processes draws the argument into a universalising direction, obliterating the need for analysing the distinctive ways in which activities and labour are embedded in complex, situated power relations and specific economic structures. In the case of Hardt/Negri's concept of an overall post-Fordist biopolitical productivity, this concept diverts attention, for example, from the ongoing difference between paid and unpaid (reproductive) labour as – although in a changing and dynamic manner – constitutive for the gendered division of work (Schultz, 2002).

Waldby and Cooper build on an intellectual lineage of materialist feminist critique, brought forward by scholars such as Margret Lock, Sarah Franklin, Charis Thompson, and Donna Dickenson, who in turn drew from an early Marxist notion of alienation and apply it to the new appropriation of female body materials like embryos or oocytes. The Marxist notion of alienation employed here refers to the relation between the subjectivity of the labourer and the results of his labour; he (and in Marx, the worker is always a 'he') is separated from the fruits of his labour, although they are nothing else than the objectified, 'dead' version of his living labour. Subjectivity turns into an alien object. Hence, alienation here is essentially conceptualised as a relation between subject and object. Yet, the problem with feminist critique building on an early Marxist notion of alienation is, we think, that alienation is basically understood as a universal relation between (body as) subject and (body as) object, not as a relation between interacting human agents. The notion of alienation, here, is tied to an ahistorical notion of labour as the metabolism of 'man' with nature, confined to a subject philosophical framework. In his later work, in contrast, Marx (1996) sketched out a notion of alienation that referred to the distorted relation between humans as social agents and the societal relations they live in. The commodity fetish as a distinctively modern, capitalist form of alienation means that the social relations in which people live, and which they have created themselves, appear as relations between things. Hence, in *Capital* it is the reification of social relations that constitutes alienation. Since human labour in capitalism has become a commodity too, the reiteration of social relations in particular applies to relations of production. However, alienation here is not a subject/object relation but a social relation, albeit one that does not appear as such. From this perspective, social critique would not be primarily about reappropriating the products of one's labour, or getting properly paid for one's labour, nor even about regaining control about the product of one's labour, but about generating different social relations based on solidarity. The point is that capitalism systematically encourages people to treat each other as a means to an end and in addition creates the appearance that this is the natural state of affairs. Ultimately, humans and human faculties function as mere means of accumulating capital, and the imperative of accumulating capital assumes the status of an end in itself.

Hence, we suggest that there is a notion of alienation in Marx that could give inspiration to feminist theory as it is derived from social critique rather than subject philosophy. How this could be spelled out in relation to oocytes for research requires further work and consideration. If, however, overcoming a socio-economic system of which instrumentalisation is the functioning principle, then commercialisation is certainly not the way to go. Paying women for supplying their eggs may, so some extent, improve the economic situation of those who have something to sell, but it reiterates and reconfirms the logic of instrumentalisation. Furthermore, it is an open question whether and to which extent an industrial relations-like bargaining model is realistic in case of oocyte provisions. There can be no universal answer to this question; the prospects of an industrial law-like approach depend entirely on the context: the social position of women, their bargaining power, collective forms of organising, the meaning of selling oocytes, whether there is a stigma attached to it or a culture of secrecy, and so on.

While this is an empirical question referring to the supply side of the oocytes for research business, another question refers to the demand side of this business: if commercialisation of oocytes forms a requirement for this strand of research to persist, as we think it does, then demand for oocytes for research is not just a given, a matter of fact, but a matter of political decision-making; it critically depends on whether commercialisation policies are adopted or not.

Feminist debate, to conclude, has made powerful interventions into the global discourse on regenerative medicine and oocytes for research, highlighting that involvement of women is indispensable in this field and forcefully insisting that oocytes are not simply raw materials 'out there', the appropriation of which is basically a technical or economic issue. Providing oocytes critically involves women, their bodies, as well as their minds, their health, their living conditions, their activity, their judgement, their life. In the following, we suggest to take the debate further, putting the emphasis less on a general analysis of the relationship between the woman and the bio-object 'oocyte' but more strongly on the interaction between research practices and their development on the one hand and its social, cultural, and institutional context, in particular regarding unequal social relations between women, notions of rights, virtues, citizenship, the structure and logic of the bioeconomy, and how it affects social relations in general.

Doing bodies: social relations and bioeconomic power structures

The contributions we now turn to examine practices of oocyte procurement as a web of social relations, focusing on social and economic relations and power structures rather than individual rights or health effects on individual women. The contributions discussed above have focused primarily on the relation between the oocyte provider and the oocyte as the biological object and less systematically on the bioeconomic complex of cloning, stem cell research, and reproductive medicine. This complex, in contrast, is at the centre of the analysis presented by So Yeon Leem and Jin Hee Park (Leem and Park, 2008), who investigate the reasons why, at the turn of the millennium, it was fairly easy to motivate South Korean women to supply oocytes for stem cell research.

They argue that in order to understand the 'egg donor culture' that manifested during the past decade in South Korea one has to take into account a paradoxical constellation that characterises the contemporary South Korean gender regime. By donating their body materials, they argue, women could achieve social visibility and position themselves as acting subjects by putting their bodies in the service of stem cell research, which has an extremely positive reputation in South Korea. According to the authors, the risks and downsides for women arising from technological developments and related body policies, ranging from intervention for purposes of population control to the cosmetics industry and cosmetic surgery, have never become an issue of concern that would have made it on political agenda in contemporary South Korean history. In a situation like this, a universalising perspective on women that sees them as powerful, active subjects is misleading. In contrast to conceptualising female subjectivity and agency in a universalising manner, they insist that 'our whole body is part of social practices that involve technologies and other peoples bodies' (Leem and Park, 2008, p. 23). If our body is part of social practices which as such are always historically distinct and context specific, there is no universal way of conceptualising women's relation to their bodies, or property in the body, or bodily productivity. Leem and Park draw on medical anthropologist Janelle Taylor, who proposed to conceptualise women's bodies in the age of biotechnology 'not only as something that individuals have but as something that people collectively do...in multiple different ways' (Taylor, 2006, p. 7). Taylor presented this view at a conference on women's rights in South Korea in the wake of the Hwang scandal. The topic was 'Envisioning the Human Rights of Women in the Age of Biotechnology and Science'. The conference was organised by the Korean NGO 'Korean WomenLink' and brought together representatives of women's organisations and feminist experts from various countries. One political issue debated was the lawsuit presented by 36 Korean women's organisations on behalf of two women who experienced serious side effects after undergoing egg retrieval for Hwang's research team, another was the involvement of women's health advocates in developing the Korean Bioethics Law (Genetic Crossroads, 2006).[19]

Understanding bodies as socially embedded practices means requires a situated analysis of the economic relations and social power relations that form the context of oocyte procurement. In our view, this includes an analysis of the current gender regime but also means to take the following dimensions into account:

First, discussing the use of oocytes in stem cell research has to scrutinise the routinisation and normalisation of 'egg harvesting' in the context of reproductive medicine and how the routines, regulatory frameworks, institutional arrangements, attitudes, expectations, and cultural imageries ruling the reproductive sector are intertwined with practices in the stem cell research sector. Close personal connections and spatial proximity between stem cell research and the IVF sector have proven critical for research strategies that require human oocytes.[20] Our research on oocyte procurement strategies in Europe has shown that but for one exception, SCNT research persisted where a tight 'IVF–stem cell interface' (Franklin, 2006) was evident and it concluded where it was lacking. The notion of 'surplus' oocytes being generated in the course of IVF treatment forms another essential link between the IVF sector and research, with 'surplus' being both a discursive construct and the outcome of a set of practices such as hormone stimulation.[21] Sometimes, we found, research projects used oocytes that had originally been supplied for IVF purposes but were later reallocated to research. Commercialising oocytes for research in one way or the other can be understood as a strategy to make research more independent from the IVF–stem cell interface in that women can be motivated to supply oocytes who are not undergoing IVF themselves. This has the additional advantage from a research point of view, that the women recruited this way are mostly much younger than the average IVF patient and their eggs are of better quality (Braun and Schultz, 2012).

In order to critically assess these connections and their implications, it is of utmost importance to not just scrutinise the resulting conflicts of interest on the level of personal goodwill and integrity but to examine the structural connections between reproductive medicine and stem cell research and the driving forces behind them. There is a structural conflict of interest between research interests in obtaining oocytes on the one hand and physicians' duty to protect the health of their patient on the other. This structural conflict cannot be solved within the confines of the contractual moment, through, for instance, postulating that potential oocyte providers should not be approached by researchers directly – since they might unduly influence women to take health risks for the sake of research – but by nurses instead. Such an approach fails to address the fact that a conflict of interest sets in much earlier: both regenerative and reproductive medicine, for different reasons, have an interest in the practice of hormone stimulation; regenerative, because there would be no available oocytes whatsoever without hormone stimulation, IVF clinics because they assume that it increases their IVF success rates and hence their profits. By now, however, IVF treatment could in fact well do without or with only minimal hormone treatment; success rates per treatment cycle may be marginally lower, but the procedure is definitely more benevolent to women's health and also much less expensive – which allows several attempts in a row without damaging the woman's health (Nargund et al., 2007; Nargund, 2009). Hence the conflict of interest cannot be adequately addressed without addressing the vested interests involved in hormone stimulation.[22]

Secondly, another issue area that merits closer scrutiny in terms of social relations embedding oocyte procurement is the trajectory of a biotechnological development overdetermined by a specific bioeconomic rationality, as Sunder Rajan has described (Sunder Rajan, 2006): stem cell research using human oocytes is an integral part of and dependent on the speculative investments of a biotech industry driven by hopes and hypes. The future on the horizon of those promises is a model of 'regenerative medicine' offering a set of personalised high-tech services, which are very likely to be restricted to private healthcare systems. In this regard, we agree with Sarah Sexton who proposed that feminists move beyond a debate revolving around individual donor rights and best practices in egg extraction, and investigate more thoroughly the potentially or already discernible socially stratifying implications of an oocyte-based research and medicine (Sexton, 2006).

Thirdly, we should come to terms with the process of piecemeal commercialisation of body materials and oocytes in particular. Our empirical research shows that 'pockets of commercialisation' have emerged in Europe in recent years, mostly in Spain and the United Kingdom.[23] All these models have developed under the auspices of regulatory regimes that prohibit the outright commercialisation of embryos and gametes 'in general'. These general non-commercialisation policies not withstanding, such practices have, however, all been sanctioned by the authorities in charge, the underlying rationale being that exceptions to the general rule are required to manage the dramatic shortage of oocytes. Given the existence of vested interests in the commercialisation of oocytes, however, the exceptions tend to turn into the norm, as we have seen in the UK case. Here, egg sharing schemes had originally been approved for IVF purposes only, although they clearly form payment in kind and as such contravened the ban on a commercialisation of gametes and embryos. In order to meet the perceived shortage of eggs for IVF purposes, the HFEA decided to make an exception to the rule. Subsequently, egg sharing for research was approved by the same authority on the grounds that like has to be treated alike and egg sharing, they reasoned, for research could not legitimately be banned when egg sharing for IVF had already been approved. Nevertheless, the scheme was constructed as forming an exception to the generally still valid non-commercialisation rule. Still a few years later, the HFEA decided, after conducting a public consultation, to adopt the Spanish model of offering 'compensations' for so-called non-patient 'donation'. Constructing these incentives as 'compensations', which by definition do not constitute 'payment', allows the government to approve de facto commercialisation while still upholding the fiction of a general non-commercialisation rule.

From the contractual moment to social critique: a shift in perspective

How could a concern with social equality and 'doing bodies' be translated into political critique? In the following, we will go back to another strand of feminist body politics that might provide an instructive model here, namely, the radical international reproductive rights movements that developed an anti-capitalist, anti-racist, and anti-eugenic critique of international population policies. We turn to these movements' lines of critique in order to further elaborate a feminist perspective on oocytes for research committed to social change and social critique. Particularly the radical reproductive rights movements' conception of rights may be helpful for this purpose. By radical reproductive rights movements, we mean the movements engaged in building international feminist networks in the 1980s out of which emerged the reproductive justice movement in the United States as well as anti-racist, anti-eugenic feminist movements against population control policies in Asia and Latin America.[24] They developed complex and diverging approaches to women's rights in the context of issues such as birth control, abortion rights, or childbearing. Feminist movements committed to social equality and social critique focused on different problems, started from different reference points and arrived at a different set of demands than liberal reproductive rights movements. For the former, a liberal, juridical, and individualistic conception of rights had little appeal (Schultz, 2007). They rather arrived at a notion of rights as collective claims that grew out of a specific dynamics of politicisation. Both this politicisation process and the notion of rights that evolved in its course might provide inspiration for feminist debate on oocyte procurement. The dynamics of politicisation can be summarised briefly in terms of (1) expanding the boundaries of problem perception, (2) performing strategies of contextualisation, (3) formulating a politically grounded critique of technology, and (4) a reflexive notion of rights as collective claims.

By expanding the boundaries of problem perception, we mean a political perspective that does not divorce individual opportunities from social and economic structures they are embedded in. In the case of reproductive rights, this means that the individual decision for or against having children cannot be separated from population policies, hegemonic family structures and gender regimes, the relations of care work, and the meaning and politics of maternity.

Placing the problem at stake in its wider context of social power relations and economic structures is closely related to strategies of contextualisation starting from the insight that the scope for individual decision-making is very different for different groups of women. The range of available options depends on women's social status and their living conditions, and on the role certain policies assign them with regard to motherhood, reproductive work and care taking which differ along the lines of class, race, able-bodiness, and other axes of difference. Faye Ginsburg and Rayna Rapp used the concept of 'stratified reproduction' to address 'the power relations by which some categories of people are empowered to nurture and reproduce, while others are disempowered'. Feminist research, they insisted, has to ask: 'who is normatively entitled to refuse child-bearing, to be a parent, to be a caretaker, to have other caretakers for their children, to give nurture or to give culture (or both)? [...]' (Ginsberg and Rapp, 1995, p. 3). In this vein, authors from the reproductive justice movement in the United States emphasised that the freedom of choice to have children or not may hinge on, for instance, whether one has the economic means to afford abortion services, but also on racist and eugenic regimes of encouragement or stigmatisation defining whose children are considered socially desirable or not (Roberts, 1997).

The reproductive rights movements, furthermore, developed a critique of technology concerned with the political and economic context of research on contraceptive and sterilisation technologies, namely, the strong financial and technological influence of population agencies such as the Population Council engaged in reducing population growth in the South. Their criticism, hence, was not derived from abstract criteria defining a 'good' contraceptive or some postulate of informed freedom of choice. Instead, they criticised the influence the global population policy establishment had on contraceptive research (Nair, 1989; Hartmann, 1995, p. 179ff.). This relationship led to the propagation of methods of birth control, particularly in countries of the South, that have a long-term impact and over which women have little control, such as implants, injections, or sterilisation (Schultz, 2006).

The political dynamics at work here gave rise to a conception of rights as collective claims. Reproductive rights became the focal point for the struggle for social change, a struggle that went way beyond a juridical framework of individual rights. Rosalind Petchesky put it this way: recourse to reproductive rights must be understood – as in other fields of human rights discourse – as an intervention into a 'discursive field of power relations' (Petchesky, 2003, p. 22). Rights 'are simply the rhetorical structure "given to us" in the present historical conditions for asserting counter-hegemonical statements of justice' (Petchesky, 2003, p. 26).

What can this line of social critique with its notion of politics and rights mean for the debate on oocytes for regenerative medicine research?

In order to answer this question, we will summarise the various shifts in focus and the elements leading to the widening and contextualisation of the debate. A feminist standpoint would go beyond issues concerned with health risks and burdens to analyse and assess the research practices using oocytes based on the bioeconomic context in which these practices are embedded and how the social relations that form this context are changing or are being newly established. This would require assessing the logic of economic speculation in the biotech industry of which cloning research is a part. We would also need to consider how 'personalised' high-tech medicine can be expected to promote social inequality should this research actually be successful some day. A critical feminist standpoint would further include an assessment of the continually expanding biotechnological appropriation of body parts and materials as well as of trends towards commercialisation, which we are currently witnessing in oocyte procurement. It would imply appraising and formulating a position regarding privatised reproductive medicine, which has evolved on a global scale and has been the institutional prerequisite and context for the appropriation of oocytes to become a routine process. In particular, the relations of social inequality that have already evolved between women as vendors and purchasers of oocytes must be addressed. Refocusing, contextualising, and widening the debate in this manner would require shifting the political speaking position from which such analyses and assessments are performed, demands raised, or rights claimed.

A feminist debate capable of transcending the boundaries outlined above would thus not speak on behalf of the interests of the individual egg donor or Waldby's imagined collective. It would move away from a perspective focused on the individuals affected and more towards the collective position of those who, for various reasons, take a critical look at the developments in reproductive medicine and stem cell research. Adopting such a change in perspective and bringing it to life, however, would require reviving the political debate on biotechnologies beyond the debates among policy advisers and academics only. Of course, this requirement points to the limitations of our proposal, which are similar to those of Waldby's collective of regenerative workers. Nevertheless, this seems the only viable path for the dynamics of social movements discussed above to gain the momentum allowing the opening up of opportunities not only for contextualising, widening, and shifting the debate concerning research on oocytes but for extending this reconfigured debate to biotechnological research and the bioeconomic logics dominating it in general. Perhaps it is no coincidence that the feminist debate on oocytes, which has mostly been led by individual experts in the past, has so far been more of an obstacle to any such shift in perspective.

Notes

1. http://www.internationalstemcell.com/news2009.htm##, accessed 18 August 2010.
2. See http://www.stemagen.com/17jan08.htm, accessed 18 August 2010.
3. In this case, women received $8000 from the Columbia University where the research was done; see (Wade, 2011). 'After setbacks in harvesting stem cells, a new approach shows Promise.' New York Times (October 5).
4. In his scientific publications, Hwang himself reported using only 185 human eggs (see Hwang et al., 2005; Hong, 2008).
5. Contributions to this debate include but are not restricted to Dickenson (2006, 2008); Beeson and Lippman (2006); Schneider (2006); Sexton (2006); Baylis and McLeod (2007); Thompson (2007); George (2008); Leem and Park (2008); Roberts (2008); Waldby (2008); Waldby and Cooper (2010); Haimes et al. (2012).
6. For health risks involved in oocyte retrieval see Delvigne and Rozenberg (2002); Hugues (2002), Magnus and Cho (2005); Norsigian (2005); Beeson and Lippman (2006); Guidice et al. (2007). A recently published study in the Netherlands concludes that hormonal ovarian stimulation may increase the risk of ovarian tumours from under 5 in 1000 women in the general population to 7 in 1000 women who had undergone ovarian stimulation, see van Leeuwen et al. (2011).
7. For more details on this scheme see Braun and Schultz (2012). Previously, the review board had decided women should not receive more than €600 *for research 'donation'*, in contrast to the €900–1000 that may be offered for reproductive 'donation' in the IVF sector.
8. See NESCI (2007, 2008); Baylis and Mcleod (2007); O'Riordan and Haran (2009).
9. For an overview over amounts paid to reproductive 'donors' in the United States see Levin (2010).
10. Similarly, Spar (2007) points at the inconsistence between reproductive and research regulations and points at proper informed consent procedures as one major condition for a good regulation of both procedures.
11. Schneider calls 'for high standards for ethical, accountable informed consent procedures' with 'full and accurate, unbiased information about short- and long-term risks of hormonal stimulation and of oöcyte extraction for a woman's health and fertility' (2006, p. 225).
12. For Locke, property emerges if and when my labour mixes with an unowned object which thereby becomes the fruit of my labour and thus my own. Hence, property rights are based on the assumption of self-ownership in the person on the one hand and on labour, action, and agency on the other. However, self-ownership and labour do not necessarily have to coincide, in Locke, in order to constitute property: 'Thus when my horse bites off some grass, my servant cuts turf, or I dig up ore, in any place where I have a right to these in common with others, the grass or turf or ore becomes my property' (Locke, 2002, p. 12). Here, the relation between me and the servant precedes the constitution of property. Hence, property in Locke does not only arise from a relation between (labouring) subject and (unowned) object but also and not least from a social relation of power and domination.

13. See NESCI (2007); Roberts and Throsby (2008); Braun and Schultz (2012); Haimes et al. (2012).

14. The ovarian hyperstimulation syndrome (OHSS) caused by hormone stimulation is characterised by ovary enlargement, a change in blood composition, and vascular hyperpermeability. Mild symptoms are nausea and increased abdominal girth. Severe complications involve renal failure, pulmonary embolism, and apoplexy. Even a small number of fatalities caused by OHSS have been reported. The risk for young women is higher than for older females. At the same time, however, the side effects of hormone stimulation are better documented for IVF patients who, on average, are older than women undergoing oocyte retrieval for research purposes only, in exchange with a monetary compensation (see Jayaprakasan et al., 2007; Balen, 2008). There is controversy over the question as to how frequently OHSS occurs in the wake of IVF procedures. The numbers reported for the incidence of severe symptoms vary between 0.1% and 5% of the women having received hormone treatment (see Braun and Schultz, 2012).

15. Gottweis and Kim (2009, 2010) argue that South Korean enthusiasm for Hwang can be understood as manifestations of a new type of nationalism they term bionationalism.

16. In earlier work, Waldby (2008) speaks of *reproductive labour*. Waldby and Cooper (2010) explicitly replace the concept or reproductive labour by the concept of *regenerative labour*. In so doing, they stress the productive potential of the oocyte, which is not least what makes it so important for regenerative medicine.

17. In the discussion following her presentation at the conference 'Regenerative Medicine in the 21st Century: Managing Uncertainty at the Global Level,' 9–10 June 2010 in Madison, WI, Waldby spoke in favour of paying egg provider.

18. See Waldby (2008) and her presentation given at the conference 'Regenerative Medicine in the 21st Century: Managing Uncertainty at the Global Level,' 9–10 June 2010 in Madison, WI.

19. See also on the programme at http://2006forum.womenlink.or.kr/abstract.php.

20. See also Dickenson and Alkorta Idiakez (2008) and for Spain in particular Alkorta Idiakez (2010).

21. On the construction of oocytes as being 'surplus' in the context of 'egg sharing' schemes see Roberts and Throsby (2008) and Waldby and Carrol (2012).

22. Another party that has vested interests in hormone stimulation is of course the pharmaceutical industry marketing the drugs.

23. We discuss these models in more detail in Braun and Schultz (2012).

24. Important organisations were Committee on Women Population and Environment, Women's Global Network for Reproductive Rights, FINRRAGE, UBINIG; important feminist intellectuals engaged in these movement were Sumati Nair, Farida Akther, Maria Mies, Rosalind Petchesky, Maria Bethania Avila, Betsy Hartmann, Jurema Werneck, and others. For the complex history of women's rights movement and their heterogeneous conceptions of reproductive rights and positions towards population policies see Schultz (2006).

References

Alkorta Idiakez, I. (2010) Egg donation: a case of body shopping. Talk given at the University of Innsbruck at 6 May 2010. Available at http://www.uibk.ac.at/fakultaeten/volkswirtschaft_und_statistik/forschung/wsg/docs/alkorta.pdf

Balen, A. H. (2008) *Ovarian Hyperstimulation Syndrome – A Short Report to the HFEA*. London: HFEA.

Baylis, F. and C. Mcleod (2007) The stem cell debate continues: the buying and selling of eggs for research, *Journal of Medical Ethics*, 33: 726–731.

Beeson, D. and A. Lippman (2006) Egg harvesting for stem cell research: medical risks and ethical problems, *Reproductive BioMedicine Online*, 13(4): 573–579.

Braun, K. and S. Schultz (2012) Oöcytes for research: inspecting the commercialisation continuum, *New Genetics and Society*, 31(2): 1–23.

Delvigne, A. and S. Rozenberg (2002) Epidemiology and prevention of ovarian hyperstimulation syndrome (OHSS): a review, *Human Reproduction Update*, 8(6): 559–577.

Dickenson, D. (2006) The Lady Vanishes: what's missing from the stem cell debate, *Journal of Bioethical Inquiry*, 3(1–2): 43–54.

Dickenson, D. (2007) *Property in the Body: Feminist Perspectives*. Cambridge, UK, Cambridge University Press.

Dickenson, D. and I. Alkorta Idiakez (2008) Ova donation for stem cell research: an international perspective, *International Journal of Feminist Approaches to Bioethics*, 1(2): 1–17.

Franklin, S. (2006) The IVF-stem cell interface, *International Journal of Surgery*, 4(2): 86–90.

French, A. J., C. A. Adams, et al. (2008) Development of human cloned blastocysts following somatic cell nuclear transfer with adult fibroblasts, *Stem Cells*, 26(2): 485–493.

Genetic Crossroads (2006) After the Hwang Scandal: Korean women's groups hold International Conference. Available at http://www.geneticsandsociety.org/article.php?id=2592

George, K. (2008) Women as collateral damage: a critique of egg harvesting for cloning research, *Women's Studies International Forum*, 31: 285–292.

Ginsberg, F. and R. Rapp (eds) (1995) *Conceiving the New World Order*. Berkeley, CA: University of California Press.

Gottweis, H. and B. Kim (2009) Bionationalism, stem cells, BSE, and Web 2.0 in South Korea: toward the reconfiguration of biopolitics, *New Genetics and Society*, 28(3): 223–239.

Gottweis, H. and B. Kim (2010) Explaining Hwang-Gate: South Korean identity politics between bionationalism and globalization, *Science, Technology and Human Values*, 35(4): 501–524.

Gottweis, H. and R. Triendl (2006) South Korean policy failure and the Hwang debacle, *Nature Biotechnology*, 24(2): 141–143.

Guidice, L. et al. (eds) (2007) *Assessing the Medical Risks of Human Oocyte Donation for Stem Cell Research: Workshop Report*. Washington, DC: National Academies Press.

Haimes, E. et al. (2012). Eggs, ethics and exploitation? Investigating women's experiences of an 'egg sharing' scheme, *Sociology of Health and Illness*, 34(8): 1199–1214.

Hands off our Ovaries (2006) July 14, 2006 – letter Regarding: California Institute of Regenerative Medicine (CIRM) Medical and Ethical Standards Regulations, 17 Cal. Code of Regs. 100010–100130. Press Releases, Op Eds, Letters. Available at http://handsoffourovaries.com/pr.htm, accessed 29 March 2009.

Hardt, M. and T. Negri (2002) *Empire: Die neue Weltordnung.* Frankfurt a.M.: Campus.

Hartmann, B. (1995) *Reproductive Rights and Wrongs: The Global Politics of Population Control. The Global Politics of Population Control. Revised Edition.* Boston: South End Press.

HFEA (2011) HFEA agrees new policies to improve sperm and egg donation services. Press release 19 October 2011. Available at http://www.hfea.gov.uk/6700.html, accessed 7 February 2012.

Hong, S. (2008) The Hwang scandal that 'shook the world of science!', *East Asian Science, Technology and Society: An International Journal (EASTS),* 2(1): 1–7.

Hugues, J.-N. (2002) Ovarian stimulation for assisted reproductive technologies. In E. Vayena, P. J. Rowe and D. P. Griffin (eds) *Current Practices and Controversies in Assisted Reproduction.* Report of a meeting on 'Medical, Ethical and Social Aspects of Assisted Reproduction' held at WHO headquarters in Geneva, Switzerland, 17–21 September 2001. Geneva: World Health Organization, pp. 102–125.

Hwang, W. S. et al. (2005) Patient-specific embryonic stem cells derived from human SCNT blastocysts, *Science,* 308(5729): 1777–1783.

Isasi, R. M. and B. M. Knoppers (2007) Monetary payments for the procurement of oocytes for stem cell research: in search of ethical and political consistency, *Stem Cell Research,* 1(1): 37–44.

Jayaprakasan, K. et al. (2007) Estimating the risks of ovarian hyperstimulation syndrome (OHSS): implications for egg donation for research, *Human Fertility,* 10(3): 183–187.

Leem, S. Y. and J. H. Park (2008) Rethinking women and their bodies in the age of biotechnology: feminist commentaries on the Hwang affair, *East Asian Science, Technology and Society: An International Journal (EASTS),* 2(1): 9–26.

Levin, A. D. (2010) Self-regulation, compensation, and the ethical recruitment of oocyte donors, *Hastings Center Report,* 40(2): 25–36.

Locke, J. (2002) *The Second Treatise of Government and A Letter Concerning Toleration.* Dover: Dover Thrift Editions.

Magnus, D. and M. K. Cho (2005) Issues in oocyte donation for stem cell research, *Science,* 2005: 308: 1747–1748. Available online at http://www.ugr.es/~perisv/congresos/lecturasfc/2005–2006/Issues%20in%20Oocyte%20Donation%20For%20StemCells%20Research.pdf, accessed 16 November 2012.

Marx, K. (1996) *Capital,* Vol. 1. London: Lawrence and Wishart.

Nahmann, M. (2008) Nodes of desire: Romanian egg sellers, 'dignity' and feminist alliances in transnational ova exchanges, *European Journal of Women's Studies,* 15(2): 65–82.

Nair, S. (1989) *Imperialism and the Control of Women's Fertility.* Arnhem: New Hormonal Contraceptives, Population Control, and the WHO.

Nargund, G. (2009) Natural/mild assisted reproduction technologies: reducing cost and increasing safety, *Women's Health,* 5(4): 359–360.

Nargund, G. et al. (2007) Low dose HGC is useful in preventing OHSS in high risk women without adversely affecting the outcome of IVF cycles, *Reproductive Biomedicine Online,* 14(6): 682–685.

NESCI (2007) Egg sharing: Women to get help with IVF treatment costs for donating eggs to research. 13 September 2007. Available at http://www.nesci.ac.uk/news/item/egg-sharing-women-to-get-help-with-ivf-treatment-costs-for-donating-eggs-to-research, accessed 7 February 2012.

NESCI (2008) North East Stem Cell Institute: Egg sharing: Successful pregnancies in world-first scheme in Newcastle. Available at http://www.nesci.ac.uk/news/, accessed 26 March 2009.

Noggle, S. et al. (2011) Human oocytes reprogram somatic cells to a pluripotent state, *Nature*, 478: 70–75, doi: 10.1038/nature10397.

Norsigian, J. (2005) Egg donation for IVF and stem cell research: time to weigh the risks to women's health. Different Takes 33. Available at http://www.global-sisterhood-network.org/content/view/227/59/

NYSTEM (2009) New York state stem cell science: statement of the Empire State Stem Cell Board on the compensation of oocyte donors. Available at http://stemcell.ny.gov/news.html, accessed 7 February 2012.

NYT (2009) New York state allows payment for egg donations for research, *New York Times*, 26 June 2009.

O'Riordan, K. and J. Haran (2009) From reproduction to research: sourcing eggs, IVF, and cloning in the UK, *Feminist Theory*, 10(2): 191–210.

Pavone, V. and F. Arias (2012) Beyond the geneticization thesis: the political economy of PGD/PGS in Spain, *Science, Technology, & Human Values*, 37(3): 235–265.

Petchesky, R. (2003) *Global Prescriptions. Gendering Health and Human Rights*. London and New York: Zed Books.

Reynolds, J. and M. Darnowsky (2006) The California Stem Cell Programme at One Year: A Progress Report Berkely, Center for Genetics and Society. Available at http://genetics.live.radicaldesigns.org/downloads/200601report.pdf

Roberts, C. and K. Throsby (2008) Paid to share: IVF patients, eggs and stem cell research, *Social Science and Medicine*, 66(1): 159–169.

Roberts, D. (1997) *Killing the Black Body: Race, Reproduction, and the Meaning of Liberty*. New York: Pantheon Books.

Schneider, I. (2006) Oocyte donation for reproduction and research cloning – the perils of commodification and the need for European and international regulation, *Law and the Human Genome*, 25: 205–241.

Schultz, S. (2002) Biopolitik und affektive Arbeit bei Hardt/Negri 44(56), *Das Argument*, 44(56): 696–708.

Schultz, S. (2006) *Hegemonie, Gouvernementalität, Biomacht: Reproduktive Risiken und die Transformation internationaler Bevölkerungspolitik*. Münster: Westfälisches Dampfboot.

Schultz, S. (2007) *Von einem expansiven Rechtsbegriff zur biopolitischen Artikulation positiven Rechts:Dimensionen der Verrechtlichung in der Geschichte der reproductive rights: Politisierung und Entpolisierung als performative Praxis. D. G. Schulze, S. Berghahn and F. O. Wolf*. Münster: Westfälisches Dampfboot, pp. 90–99.

Schultz, S. (2010) Women's Eggs for Research: Without Payment? Center for Genetics and Society, 14 January 2010. Available at http://www.geneticsandsociety.org/article.php?id= 5413

Sexton, S. (2006) Ethics of Economics? Health or Wealth? Beyond Ova in the Lab. Paper presented at the Conference: 'Envisioning the Human Rights of Women in the Age of Biotechnology and Science', Seoul, South Korea, 21 September 2006.

Steinbock, B. (2009) 'Paying egg donors for research: in defense of New York's landmark decision, 07/01/2009', *Bioethics Forum*. Available at http://www.thehastingscenter.org/Bioethicsforum/Post.aspx?id= 3638

Stojkovic, M. et al. (2005) Derivation of a human blastocyst after heterologous nuclear transfer to donated oocytes, *Reproductive BioMedicine Online*, 11(2): 226–231.

Sunder Rajan, K. (2006) *Biocapital: The Constitution of Postgenomic Life*. Durham and London: Duke University Press.

Taylor, J. S. (2006) 'Biotechnology' and 'Women's Bodies': Hazardous Concepts? Unpublished manuscript, prepared for the international 'Envisioning the Human Rights of Women in the Age of Biotechnology and Science', in Seoul, South Korea on 20–21 September 2006.

Thompson, C. (2007) Why we should, in fact, pay for egg donation, *Regenerative Medicine*, 2(2): 203–209.

Tsuge, A. and H. Hong (2011) Reconsidering ethical issues about 'voluntary egg donors' in Hwang's case in global context, *New Genetics and Society*, 27(3): 241–252.

van Leeuwen, E. E. et al. (2011) Risk of borderline and invasive ovarian tumours after ovarian stimulation for in vitro fertilization in a large Dutch cohort, *Human Reproduction*, 26(12): 3456–3465.

Wade, N. (2011) After setbacks in harvesting stem cells, a new approach shows promise, *New York Times*, October 5.

Waldby, C. (2008) Oocyte markets: women's reproductive work in embryonic stem cell research, *New Genetics and Society*, 27(1): 19–31.

Waldby, C. and K. Carrol (2012) Informed consent and fresh egg donation for stem cell research, *Journal of Bioethical Inquiry*, 1(1), doi: 10.1007/s11673-011-9349-4.

Waldby, C. and M. Cooper (2010) From reproductive work to regenerative labour: the female body and the stem cell industries, *Feminist Theory*, 11(3): 3–22.

Waldby, C. and R. C. Mitchell (2006) *Tissue Economies: Blood, Organs and Cell Lines in Late Capitalism*. Durham, NC: Duke University Press.

6
Cloning and the Oviedo Convention: The Socio-cultural Construction of Regulation

Itziar Alkorta, Inigo Miguel Beriain, and David Rodríguez-Arias

Introduction

In 1997, the Oviedo Convention, signed by most of the European Union (EU) member states (Germany and the United Kingdom were the most remarkable exceptions), banned 'the creation of human embryos for research purposes'. At that time, this ban did not seem to be decisive for the future development of regenerative medicine (RM). It was only two months prior to the scheduled date for the signature of the Convention that the birth of 'Dolly' was announced to the world and therapeutic applications of somatic cell nuclear transfer (SCNT, discussed previously in Chapter 5) had only just started to be envisioned.

In the years that followed, the biomedical field was to change dramatically: therapeutic cloning attracted major attention as a promising source of allogeneic stem cells which could be used for research into cell therapies. RM was claimed by scientists working in this field as a paradigmatic shift in the meaning and practice of medicine itself (see Chapter 9). But doing research with embryos now posed a major problem to countries like France, Finland, or Spain, which had just ratified/signed the Oviedo Convention, and at the same time, were willing and keen to participate in the growing competition between as well as collaboration among the members of international stem cell research community. This chapter addresses the way these countries faced that challenge, pointing out to the strategies followed by them to permit the development of the field at a corporate level, while preserving the normative limits of embryo research imposed by the Oviedo Convention. It shows how modifications in the definition of the term 'embryo' made it possible to formally keep the ban on embryo creation for research purposes while allowing SCNT research to go ahead within these countries in the wider international and indeed global arenas.

In this sense, the Oviedo Convention can be presented as a boundary object (Star and Griesemer, 1989), serving the needs of those countries needing to respond to the ethical legitimacy debate that embryo research had provoked in their communities. Initially conceived to ban the creation of research embryos (see Chapter 7), in practice, however, the Convention has been used by some countries to provide ethical legitimation for the creation of human embryonic stem cell lines (hESC). This chapter argues, paradoxically, that the Oviedo Convention has played a major role in the social acceptance of RM as a new medical paradigm across Europe, mobilising the field across different legal *and* ethical boundaries.

Who needs a convention?

Increasing globalised research and world scientific collaboration pose challenges to national regulation in terms of the interoperability, harmonisation, and convergence of international regulations on RM in a context of cultural diversity. A fragmented regulatory landscape, as is the case within Europe, has proved to be extremely inefficient from the point of view of transnational research operability and effective allocation of resources, notwithstanding the moves towards the standardisation of stem cell experimental techniques outlined in Chapter 1. In particular, variation and diversity on hESC and cloning national regulations are especially pronounced and much more intense and contested than in other fields of health technology. Within Europe, each national legal system establishes a specific regulatory framework regarding the derivation and use of stem cells. Indeed, this is echoed in the United States where, although there is a single regulatory framework for RM under the aegis of the Food and Drug Administration (FDA), federal and state policies also diverge in regard to what is permitted and why. In response to this, a degree of legal and regulatory harmonisation has been sought for many years by scientists, clinicians, and corporations, the latter keen to establish a common ground on which their products receive market authorisation.

This is not easy to deliver however, since the global development of RM and its therapeutic applications is largely dependent on the nature of its engagement (at national and international levels) with key religious and cultural values and beliefs regarding, for instance, the moral and legal status of human tissue, the sources of oocytes (discussed more fully in Chapter 5), the risks of clinical trials, and the inequalities produced by tissue exchange dynamics. In particular, the moral and legal status of human embryos has been a major issue in many countries within continental Europe, when and where policymakers have been confronted by the demands of scientists for the authorisation of SCNT. In this context, states such as France or Spain – which hosted the process of negotiation and the signing of the Oviedo Convention – soon looked for help from the Council of Europe (CoE).

In line with its Statute, the CoE seeks to protect the individual's dignity and fundamental rights with regard to the impact of the applications of biology and medicine, what, as was noted earlier in Chapter 1, brings to the fore what Jasanoff (2011) calls the challenges of bioconstitutionalism. The CoE, even though it acknowledged that stem cells could be an important source of progress for human health and quality of life, recognised that these developments raised new concerns with regard to the protection of human dignity and fundamental rights and freedoms. To that end, the CoE, through its Steering Committee on Bioethics (CDBI)[1] worked at defining principles and establishing legal standards which would be internationally applicable across all its member states.[2]

The *Convention for the Protection of Human Rights and Dignity of the Human Being with regard to the Application of Biology and Medicine: Convention on Human Rights and Biomedicine* (commonly known as the Oviedo Convention), the first European legally binding instrument in the field, was meant to provide a framework for the protection of human rights and human dignity by establishing fundamental principles applicable to daily medicine as well as to new technologies in the fields of biology and medicine. Later additional protocols to the Convention developed these principles in greater detail in specific fields such as cloning, human organ and tissue transplantation, biomedical research, and genetic testing for health purposes.

The Oviedo Convention regulation on SCNT

The Oviedo Convention regulates on a number of areas related to biomedicine – some of which are not strictly related to stem cell research. This includes the protection of research subjects, informed consent, advanced care planning, organ and tissue transplantation, commodification of human body parts, and governance. The Convention prohibits the creation of human embryos for research purposes. This provision does not preclude research on human embryos, but limits it to the 'surplus' (or what are called 'supernumerary') embryos from reproductive attempts in a process of in vitro fertilisation (IVF) therapy.

'Spare' embryos have the genetic information of two individuals who are genetically distant from the patients likely to benefit from a stem cell-based therapy. This circumstance involves a therapeutic limit due to problems of incompatibility and the risk of rejection through the immune response. Compared to stem cells obtained from supernumerary embryos, SCNT offers a more specific and advantageous source for (personalised) RM since the core genetic material is derived from the cells of the patient needing the treatment.

Research involving human embryos promises exciting therapeutic advances, but it also raises ethical and moral dilemmas for scientists and potential donors. The most feared – and globally prohibited outcome – is to create human clones. Reproductive cloning differs from 'therapeutic cloning' in that the cloned embryo resulting from SCNT is successfully implanted and developed in a human uterus (or a potential surrogate device). As a result, despite its being driven by therapeutic and not reproductive intent, SCNT has been the source of the most controversial aspect of the whole regulation.

In fact, the lack of unity among the European countries in their endorsement to the Oviedo Convention can be attributed to this particular aspect of the convention. While some countries (e.g., Germany) have not signed the Convention because it was deemed too tolerant – in that it allows some types of embryo research – others (e.g., the United Kingdom) have refused to sign the Convention because it was considered too restrictive in that it does not give researchers enough freedom to do research with human embryonic stem cells (hESCs). The majority of European countries, including Portugal, Spain, Austria, Norway, Finland, Denmark, Romania, Bulgaria, Greece, Estonia, Lithuania, and Slovenia – have signed and ratified the convention (see Figure 6.1). By ratifying the convention, most of these countries have ruled out the possibility of creating embryos for research purposes – that is, 'therapeutic cloning' – in their territory. Interestingly, Spain, one of the countries that most firmly expressed its endorsement to the convention, is an exception in this respect. Finland is another one. Both countries have ratified the Oviedo Convention yet allow SCNT. How has this been possible?

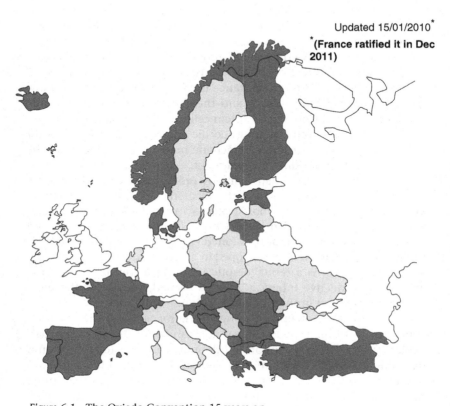

Figure 6.1   The Oviedo Convention 15 years on
Source: http://www.coe.int/t/dg3/healthbioethic/activities/01_oviedo%20convention/map_
en.asp

The creation of human embryos for research purposes is prohibited[3]

According to the letter of the Oviedo Convention, all those countries that ratified this document would have to develop regulations forbidding SCNT and, therefore, the creation of human clones regardless of the (therapeutic) purposes of such a procedure.[4] Fifteen years after the signature of the Oviedo Convention, Europe continues to be divided into three main blocks of countries. Biomedical regulations in the majority of European countries have indeed fulfilled this expectation, but not Finland and Spain (see Figure 6.2).[5]

If parallel to this we study the evolution of research with stem cells resulting from nuclear transfer, we will soon realise that the Oviedo Convention has only been a serious obstacle to a minimum number of countries. In some such as Germany and Poland, it has not entailed any problem, as they neither signed the Convention nor attempted to develop this type of research. Neither was it a problem for the Netherlands, Italy, or France, insofar as they did not develop research based on nuclear transfer even though they had signed the Convention. For its part, the United Kingdom *did* allow this type of research although, given the fact that it had never signed the Convention, it did not need to deal with any legislative challenge. Lastly, in the case of Sweden, the fact that the Convention was not ratified enabled its Parliament to pass some regulations that permitted research with stem cells created via SCNT.[6]

In short, we find that there are just two countries, Finland and Spain, which had to deal in practice with the problems arising from trying to make the political will to foster nuclear transfer for research purposes compatible with the existence of a Convention that had been signed and ratified by their respective Parliaments. Both resolved these difficulties in similar ways by sharing a common idea: a scrupulous respect for the letter of the Convention and a violation to a large extent of its spirit.

The Finnish solution

In Finland, the *Medical Research Act*, passed in 1999, introduced the following text, which has now been repealed[7]:

Section 26: Unlawful intervention on the genome

Any person who conducts research with the aim of:

1. cloning human beings;

2. creating a human being by combining embryos;

3. creating a human being by combining human gametes and genes from animals

Shall be fined or imprisoned for a period not exceeding two years for unlawful intervention on the genome.[8]

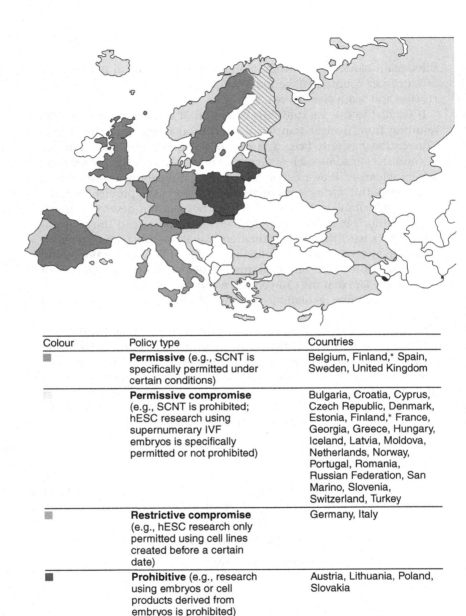

Colour	Policy type	Countries
■	**Permissive** (e.g., SCNT is specifically permitted under certain conditions)	Belgium, Finland,* Spain, Sweden, United Kingdom
▨	**Permissive compromise** (e.g., SCNT is prohibited; hESC research using supernumerary IVF embryos is specifically permitted or not prohibited)	Bulgaria, Croatia, Cyprus, Czech Republic, Denmark, Estonia, Finland,* France, Georgia, Greece, Hungary, Iceland, Latvia, Moldova, Netherlands, Norway, Portugal, Romania, Russian Federation, San Marino, Slovenia, Switzerland, Turkey
■	**Restrictive compromise** (e.g., hESC research only permitted using cell lines created before a certain date)	Germany, Italy
■	**Prohibitive** (e.g., research using embryos or cell products derived from embryos is prohibited)	Austria, Lithuania, Poland, Slovakia

Figure 6.2 Oviedo convention countries and regulatory positions

*Finland is categorised with green and yellow stripes because the relevant law (The Act on Medical Research – No. 488/1999) does not consider the product of SCNT to be an embryo. This law explicitly allows for the use of supernumerary embryos for hESC research and it is understood that SCNT – as it is not prohibited – is permitted in the country.

Thanks to the inclusion of this section, Finland was able to claim that it was strictly complying with the requirements that were stipulated in section 18.2 of the Oviedo Convention. However, how can this penalty be made compatible with the fostering of research into cells obtained via SCNT? The formula used was a relatively simple one: a definition of a human embryo was used that excluded the fact that cells obtained via this procedure would be considered as being human as such. Thus, in this Act, it is stated that 'embryo means a living group of cells resulting from fertilization not implanted in a woman's body'.[9]

In this way, although section 26 quoted above established penalties for the creation of embryos for research purposes, these were not extended to biological structures originating from nuclear transfer which, by *not* deriving from fertilisation, did not fall under the definition of 'embryo'. Regulatory conflict was thus avoided: Finnish legislation penalised the creation of embryos for research purposes as the Convention demanded, but avoided blocking the use of cells deriving from SCNT as it did not consider these to be embryos. The possibility – included in the Convention – that each country might define the embryo as it saw fit therefore proved to be of great help for such purposes.

The Spanish case

The Oviedo Convention came into effect in Spain on 1 January 2000, following its ratification by Parliament. At that time, the *Partido Popular* was in power – a conservative party which, being strongly influenced by pressure groups close to the Catholic Church, had blocked research into cells deriving from nuclear transfer. However, this situation would change significantly some years later. The socialist party Partido Socialista Obrero Español (PSOE) won the March 2004 elections, the most tumultuous in modern Spanish democracy.[10]

The context for the field of RM in general and for that of research into stem cells would prove particularly complex for the new Spanish government, because of two main factors. The first of these related to the activism of pressure groups representing different interests, each of which advocated a modification in the regulation that was adjusted to its own aims. Firstly, a group made up of research scientists themselves,[11] who had already pressurised the former government in order to obtain a regulatory modification that would bestow legal security on the experiments that were then being carried out, as well as in order to amplify the range of those that could be carried out, especially to cover stem cells produced by means of nuclear transfer. In this group the pressure exerted by Bernat Soria himself, who expressed in public his intention to move his research, from Singapore to Spain only if he were offered legal guarantees, should be noted.[12]

Secondly, the group consisting of patient associations played an important role too. In this case, diabetic associations were the most belligerent, spearheading the fight for regulatory acceptance of experiments with embryonic stem cells. So, for example, in February 2002, the Federation of Spanish Diabetics (Federación de Diabéticos Españoles) delivered to the Ombudsman 1,330,000 signatures in favour of research using embryonic stem cells.[13] This second group also strongly supported the use of hESC for research purposes. Contrary to both the scientific and patient lobbies, there was a group composed of associations for the defence of human life, closely linked to the Catholic Church, with strong support in Spain. These associations were, in turn, backed by some denominational universities (the University of Navarre, Francisco of Vitoria, etc.), professional or academic associations (like the Spanish Association of Bioethics and Medical Ethics – AEBI), as well as multiple intellectuals and university professors. All these associations were strongly opposed to any type of regulatory expansion which would involve the creation of embryos for research. This implied an emphatic no to the use of the nuclear transfer technique with human cells for the generation of stem cells.

The second key factor that the Spanish government had to face was a complex regulatory field. As mentioned, the Oviedo Convention was a strong obstacle to the use of hESC. However, it was not the only one. The 1978 Spanish Constitution indicated in its article 15 that 'Everyone has the right to life and physical or moral integrity, without in any case being submitted neither to torture nor inhuman or degrading punishments or treatments.' The interpretation that should be given to the expression 'everyone' had been the object of great controversy starting from the promulgation of the first Spanish laws regarding the decriminalisation of abortion in specific cases in 1985.[14] At that time, in its famous statement '53/1985', the Spanish Constitutional Court had already declared that embryos should not be considered people, but merely as entities worthy of protection, at least within reason,[15] which meant as much as denying them all possible tenure of the fundamental right to life. Going a step further, the Constitutional Court pronouncements 212/1996 and 116/1999 had rejected their condition as persons in legal terms.[16] Nevertheless, those same sentences adopted a gradualist approach regarding the protection to be given to human life which excluded the consideration of anything submitted to commerce, the latter indicating in particular a broad opposition to use human life/tissue as merely a means to a (corporate) end.

The final result of all this jurisprudential development was the construction of a regulatory framework regarding human embryos based on two fundamental issues. Firstly, consideration was given to differentiating between viable and non-viable embryos. If the former had the privilege of being regarded as protected legal goods, the second, characterised as such according to only biological criteria, were not susceptible to this same condition, being considered in practice as an entity equivalent to any other human biological structure. Secondly, a distinction between the creation of embryos for research and the use for such a purpose of embryos left over from assisted reproduction technology was implicitly established. While the second was explicitly accepted, the first seemed to clash radically with the consideration of legal rights protected by the Spanish Constitution.[17]

The Spanish government and its circumstances: possible options

The support for research with stem cells was, from the beginning, one of the firm policy lines of the new government. More specifically, the state executive contemplated the possibility of protecting with regulations the constitution of cellular lines by means of nuclear transfer. This policy confronted two hurdles of great significance: on the one hand, the opposition of the 'prolife' movements; on the other, the regulatory obstacles discussed above. To address these obstacles, the government had several options:

- Not to act, that is to say: leave a regulatory void with regard to nuclear transfer.
- Reject the Oviedo Convention in order to permit research with human embryos created for that purpose.
- Attempt to find some legal formula which would allow them to approve nuclear transfer without explicitly admitting the creation of embryos for research purposes without having to reject the Convention, following the Finnish path.

Following any one of these options would have different consequences and implications as we now go on to discuss.

The possibility of not acting

The first option is quite easy to describe: the Spanish government could have done nothing. Since it was quite unclear what could be considered an 'embryo' according to the Spanish laws, it may happen that, in practice, hESC research could have been allowed to develop by itself. However, this strategy presented several problems. The first was that a regulatory void was not an ideal solution for the scientific community that would then have to carry out its work with a high degree of uncertainty as to the regulatory protection it needed both within Spain and in regard to its international credentials and collaboration across the global stem cell research networks: this has been particularly true in stem cell banking where the Spanish and UK Stem Cell Banks have established strong collaborative links. Neither did it seem that a solution of this type would eliminate the response from pressure groups averse to SCNT, who could make themselves easily heard if a scientist were to put into practice this type of experiment.[18] To this it must be added that the intention to develop a broader 'Law of Biomedical Investigation' in fact excluded this possibility, given that it was clearly impossible to ignore this issue in a regulation of that type. In short, these factors meant it was impossible for the Spanish government to have recourse to this option.

The possibility of amending or rejecting the Oviedo

The second option that the Spanish executive government could adopt would be just to present an amendment to the Convention, a possibility anticipated in its articles 5 and 6. Nevertheless, the procedure of modifying the document was so exceedingly slow and complex that it would only with difficulty be reconcilable given the compelling need to approve the Law of Biomedical Investigation. Furthermore, nothing guaranteed that the rest of the signatory countries would accept the introduction of such an amendment. The sum of all of these determinants made it, in practice, exceedingly arduous to embark on that road.

A simpler option from a procedural point of view could have been the rejection of the Oviedo Convention. Now, this possibility was complex for several reasons. Firstly, the time necessary for the presentation of reserves had already gone by when the possibility of developing the Law of Biomedical Investigation arose.[19] Secondly, withdrawing from an agreement signed so recently did not seem overly serious. This was particularly important if we keep in mind that the Convention had been signed in Spain itself, which had given the country a special prominence in and at least a public sense of support for its proceedings. To this it must be added that said denounce would not directly provide a green light for the use of cellular structures obtained by nuclear transfer, but rather would leave this issue in the hands of a decision by the Constitutional Court, if anyone were to take the question to that forum. The withdrawal from the agreement would bring with it, therefore, serious inconveniences in exchange for very few advantages.

The search for an acceptable legal formula

In light of the other options, it seemed evident that the best alternative for the government would be to attempt to find some legal formula which would allow them to approve nuclear transfer without explicitly admitting the creation of embryos for research purposes or having to denounce the Convention. This complex formula would probably allow them to satisfy the interests of the different pressure groups as well as the legal limits to which the executive government found itself submitted. But was this possible from a technical point of view? As we will see in the following discussion, Spain demonstrated that it was indeed.

The Spanish solution

The solution finally adopted by Spain was that of considering, simply, that a nuclear transfer under no circumstances generates a human embryo. For this purpose, Law 14/2007, of 2 July, on Biomedical Research defined an embryo (article 3.1) as 'a phase of embryonic development from the moment in which the fertilised oöcyte is found in the uterus of a woman until the beginning of organogenesis and which ends 56 days from the moment of fertilization, with the exception of the computation of those days in which the development could have been stopped'.

In this way, the regulation designated as embryos only the cellular structures which come into being as the result of a process of fertilisation. Given that in a nuclear transfer there is no fertilisation whatsoever, it is obvious that in that case we cannot speak of a human embryo. In this way, the 'Spanish solution' imitated to a large extent the 'Finnish solution' (in both cases the key piece of the puzzle consisted in maintaining a definition of embryo strictly linked to fertilisation), but solving the legal void that was produced in the Nordic country (in Spain, unlike Finland, there did exist an explicit regulatory precaution that facilitated the use of cellular nucleus transfer techniques) much more satisfactorily.

In concordance with this approach, article 33 of the Law indicates the following:

Article 33. Obtaining of embryonic cells.

1. The creation of human pre-embryos and embryos exclusively for experimentation purposes is prohibited.

2. The use of any technique for obtaining human stem cells for therapeutic or research purposes is allowed, but only when it does not entail the creation of a pre-embryo or an embryo exclusively for this purpose, in accordance with the terms provided in this Law, including the activation of ovocytes through nuclear transfer.

Based on this terminological construction, Spanish legislators affirmed in the Preamble to the Law that 'in accordance with the gradualist perspective on the protection of human life set out by our Constitutional Court in rulings such as 53/1985, 212/1996 and 116/1999, this Law expressly prohibits the creation of human pre-embryos and embryos exclusively for the purpose of experimentation. However, the use of any technique for collecting embryonic stem cells for therapeutic or research purposes that does not entail the creation of a pre-embryo or of an embryo exclusively for this purpose, and in the terms provided by this Law, is allowable.'

The outcome of this solution was that all the pressure groups showed themselves to be sufficiently satisfied for the law to be accepted without any major explicit controversies. The acceptance of research by means of nuclear transfer satisfied the majority of patient associations or the researchers and academics that supported them. The inclusion in the law of an article whose wording explicitly prohibited the creation of embryos for research, as far as they were concerned, allowed the pro-life movements in general and the Catholic Church to save themselves the trouble of organising opposition against a law that had a great deal of support from Spanish citizens. This, in turn, relieved the *Partido Popular* of all pressure to present an appeal of unconstitutionality before the law. This risk of appeals related to the provision being unconstitutional was minimised in the law's preamble itself, explaining with a great amount of detail just why it was congruent with previous regulations.

Criticism of the Finnish/Spanish solution

So, both Finland and Spain were able to create a regulatory and legal framework that enabled SCNT research while still presenting themselves as supporters of the Convention. However, debate continues in bioethical and legal circles about the approach adopted by the two countries. Some authors, even without making a specific mention to these concrete cases, have accepted the argument that supports them.[20] Others, however, have indicated that the consensus apparently secured comes from a 'label fraud'[21] consisting in having taken away the name of embryo from a biological entity that, in reality, has, at least sometimes, the same potential to become a human person as a zygote has from a process of fertilisation. Does this type of criticism make any sense? On the one hand, it may seem that it does. We should not forget that the transfer of cell nuclei has been capable of generating adult individuals in other species of mammals, Dolly the sheep being only the first example of this kind. This evidence has led several European countries[22] and Japan[23] to change the traditional definition of an embryo as being the product of fertilisation to another in which the decisive element in the definition is *the potential for a biological structure.* Therefore, if the transfer of a nucleus can create a human embryo, we could conclude that an omission such as the one contained in the Spanish legislation does not adequately fit with the current scientific-regulatory situation.

At the same time, concluding that the transfer of the nucleus inevitably creates a human embryo is doubtful from a scientific point of view.[24] On this point, it is necessary to bear in mind that no one up until now has been able to create a human biological structure through the transfer of a nucleus that is able to be implanted in a woman's uterus. Nor, of course, has anyone been able to clone a human being, even though a legal vacuum continues to exist in many countries, allowing attempts to do so (most controversially by the Italian gynaecologist and clinical researcher Severino Antinori in 2003, who claimed to have used cloning to induce pregnancy in three women). From the legal point of view, furthermore, this statement is neither endorsed by the Oviedo Convention, its Additional Protocol, or the latter's Explanatory Report. Instead, it seems to contradict what can be concluded from them.[25]

In short, the most reliable conclusion would probably be to maintain that from a scientific point of view it would be absurd to argue that all transfers of nuclei give rise to an embryo as they never do, at least if one accepts the notion of an embryo being based on the idea of the potential to create a person. This is the most widely accepted view today in legal doctrine and in some of the legal systems that have already accepted the changes that have occurred in biotechnology previously discussed.[26]

It is doubtful, therefore, that the Spanish Law on Biomedical Research or the Finnish Medical Research Act are adequate for the current state of knowledge regarding embryos. It is possible, nevertheless, that this lack of precision is deliberate, accepted as the cost that had to be paid to satisfy the demands of all the pressure groups involved in the discussion.

The possible creation of a human clone in the future would probably make this form of definition unsustainable, but there are two points that have to be made in its defence. Firstly, in the case of Spain, it should be highlighted that the Law on Biomedical Research is not trying to define the embryo in general, but rather to define it only 'for the purposes of that law'. This can, of course, lead to legal inconsistencies (it does not make sense that in the same legal system a particular entity is defined in different ways according to what suits the legislator). Secondly, it has to be borne in mind that both the Spanish Law on Biomedical Research and the Finnish Medical Research Act will surely manage to cover SCNT in humans in sufficient time for this technology not to be necessary. The emergence of new options such as induced pluripotent stem cells or even the new possibilities that emerged from the direct use of adult stem cells seem to indicate this. Consequently, this provisional solution will have had an optimum result, if we take into account all the determining factors. Still, the question will always remain: does it make sense for a country to sign a Convention if it is willing to find a way to avoid its provisions whenever needed?

Conclusion

This chapter has explored the ways in which regulatory constraints associated with major international conventions, here the Oviedo Convention, can be worked around but also, somewhat paradoxically, actually mobilised by state agencies seeking to support and exploit research within their existing jurisdictional provisions. At the same time it is clear that the legal principles upon which regulation is based can change over time, as we saw in regard to the moves away from the definition of an embryo based on fertilisation to one based on a viable biological structure. This relation between principle and regulatory practice is indicative of the play and counter-play of science, the state and public and private interests, a theme which is explored more fully in Chapter 8. We also saw in Chapter 4 how the governance of the field of stem cells and the wider RM domain is characterised by uncertainties which are difficult to resolve and stabilise. It is also important to note that while a country, such as Germany, may have a quite restrictive regulatory regime, it has in effect offset that self-imposed constraint by allowing the importation of cell lines from countries with a more permissive regime. This global movement of cell lines that circumvent limitations at the local level can be very significant: Germany is, as Chapter 3 shows, one of the stronger centres for RM in Europe.

What this chapter also shows is how formal regulation, such as the Oviedo Convention, can act as a boundary object enabling different and at times divergent interests to be served while still allowing stakeholders a seat at the bioethical table. The bioethical standards enshrined in the convention not only are open to local interpretive process but over time are superseded by or come into conflict with other interests and other forms of regulation driven more by a scientific and technical risk assessment process (as in, for example, the regulations used to classify RM products) than one driven by a desire to manage cultural risk.

Notes

1. The CDBI is composed of representatives of 47 member states of the CoE; the Parliamentary Assembly of the CoE which has been behind many of the organisation's major initiatives; international organisations active in the field of bioethics – in particular the EU, the Organisation for Economic Cooperation and Development (OECD), the World Health Organisation (WHO), and United Nations Educational, Scientific, and Cultural Organisation (UNESCO); non-member states such as Australia, Israel, Canada, Mexico, and the United States of America.

2. Over the years the CoE has developed a network of experts comprising scientists, medical doctors, lawyers, and philosophers. In this context, a substantial set of legal instruments had already been adopted by the CoE and served as a reference point in the field of bioethics at the international level: Recommendation R (93) 4 of the Committee of Ministers to member states concerning clinical trials involving the use of components and fractionated products derived from human blood or plasma; Recommendation R (92) 3 of the Committee of Ministers to member states on genetic testing and screening for health care purposes; Recommendation R (92) 1 of the Committee of Ministers to member states on the use of analysis of deoxyribonucleic acid (DNA) within the framework of the criminal justice system; Recommendation R (90) 13 of the Committee of Ministers to member states on prenatal genetic screening, prenatal genetic diagnosis, and associated genetic counselling; Recommendation R (90) 3 of the Committee of Ministers to member states concerning medical research on human beings; Recommendation R (94) 1 of the Committee of Ministers to member states on human tissue banks, and so on.

3. Convention for the Protection of Human Rights and Dignity of the Human Being with regard to the Application of Biology and Medicine: Convention on Human Rights and Biomedicine. Oviedo, 4.IV.1997, art. 18, Research on embryos in vitro.

4. The Additional Protocol to the Convention for the Protection of Human Rights and Dignity of the Human Being with regard to the Application of Biology and Medicine, on the Prohibition of Cloning Human Beings clearly

forbids human cloning: 'Any intervention seeking to create a human being genetically identical to another human being, whether living or dead, is prohibited.'

5. See http://www.hinxtongroup.org/wp_eu_map.html
6. In that sense, the Swedish Code of Statutes No. 2006: 351, entitled The Genetic Integrity Act (2006, p. 351), dated 18 May 2006, states in its *Chapter 5. Measures for Purposes of Research or Treatment Using Human Eggs,* section 3, that

> Experiments for the purpose of research or treatment on fertilised eggs and eggs used for somatic cell nuclear transfer may be carried out no longer than up to and including the fourteenth day after fertilisation or cell nuclear transfer respectively. If a fertilised egg or an egg used for somatic cell nuclear transfer has been used for such an experiment, it shall be destroyed without delay when the measure has been accomplished.

7. Sections 25–26 were repealed by Act 375/2009.
8. See http://97918.livecity.com/97918/Finland2
9. See section 2, devoted to definitions, http://www.finlex.fi/en/laki/kaannokset/1999/en19990488.pdf
10. They took place some days after the terrorist attacks of 11 March in Madrid.
11. This stance had been reinforced some years before by a large number of Spanish universities (34), which ended up sending letters addressed to the then Spanish Minister of Science and Technology, Josep Piqué, for him to authorise experiments with embryonic stem cells by means of the use of available frozen embryos or those which can be generated for their use in cell therapy (see http://www.elpais.com/articulo/sociedad/34/universidades/piden/Pique/permita/investigar/celulas/madre/elpepisoc/20021026elpepisoc_3/Tes).
12. In an interview given to the newspaper *El Mundo* in February 2003, the scientist was asked about the offer that the autonomous government of Andalusia had made him, about taking his research with embryonic progenitor cells, to which he responded:

> I have answered that if I had said yes to Singapore I wasn't going to say no to Seville. I comprehend that the autonomous governments have many rivalries regarding public health material. The thing is that I want a written offer, in which I am authorised to work with embryonic progenitor cells, accompanied by a well-reasoned legal report. The frustrating thing would be to begin the project, obtain the funding for said project, and then, later, resign because we cannot start. It's the hardest thing that can happen to a scientist.
>
> (http://www.elmundo.es/salud/2003/513/1044636161.html)

13. http://hazte-escuchar.blogalia.com/historias/2964
14. Regarding the legal statute of human embryos in Spanish regulations, see Femenía López, P. J., *Status jurídico civil del embrión humano, con especial consideración al concebido in vitro,* Madrid: McGraw-Hill, 1999.

15. The life of the nasciturus, as soon as it represents a fundamental value – human life – guaranteed in article 15 of the Constitution, represents a legal right whose protection is found in said fundamental constitutional precept.
16. 'It should be remembered that neither non-implanted embryos, nor, indeed, mere gametes are, for all intents and purposes "human persons", therefore their availability for the Banks following the course of the fixed amount of time, can only with difficulty go against the right to life (article 15 of the Spanish Constitution) or human dignity (article 10.1 of the Spanish Constitution)' (STC 116/1999, f. j. no. 11).
17. In fact, the LIB (the Law of Biomedical Research) declares in its preamble that 'in accordance with the gradualist perspective on the protection of human life set out by our Constitutional Court in rulings such as 53/1985, 212/1996 and 116/1999, this Law expressly prohibits the creation of human preembryos and embryos exclusively for the purpose of experimentation'.
18. Even if reporting it to the courts. Let us not forget, in this sense, that cloning is a crime in Spain.
19. See article 36 of the Convention.
20. In fact, this line of argument was already assumed by a judge of the British High Court, Justice Crane. See Plomer (2002). There are, furthermore, authors of recognised prestige that have supported this same hypothesis. See also Savulescu (1999, p. 90); Atlan (1999, pp. 36, 37).
21. Romeo Casabona has written on this matter that 'the proposals which attempt to mark the differences between one zygote and another, looking for their own new designation for that which is obtained by means of activation of oocytes by nuclear transfer, are not acceptable, as that contributes nothing significant to distinguish or distance one reality from another. In Law, this resource is known as "label fraud": through a certain *nomen iuris* one tries to mask a designated situation or legal treatment that has nothing to do with its formal designation' (cf. Romeo Casabona, pp. 90 y 91).
22. In Germany, the Law guaranteeing the protection of embryos with regard to the importation and use of embryonic stem cells of human origin (Stem Cell Law), of 28 June 2002, offers, in its article § 3.4 the following definition of embryo: 'An embryo is any human totipotent cell that has the ability to divide itself and lead to a human individual as long as the necessary conditions required for said process are fulfilled.' In this same sense, it would be convenient to mention the Belgian regulations, which define an embryo as the 'cell or united system of cells with the ability to develop and lead to the growth of a human being'. In an almost identical sense, the Dutch Embryo Act of 1 September 2002 indicates, in its first section, dedicated to definitions, that an embryo is 'the cell or cell group with the ability to develop and lead to the growth of a human being'.
23. In Japan, the Law regarding regulations about techniques of human cloning and other similar techniques, of 30 November 2000, characterises, in its article 2, an embryo as 'a cell (excepting germ cells) or cells that could become a human being by means of *in utero* development in a human or animal, and that has/have not yet begun the forming of the placenta'.
24. However, it could be argued that it has never been done because it is legally banned, not because of scientific incapacity.

25. See, in relation to this, Romeo Casabona (pp. 108, 109).
26. See De Miguel Beriain (2008).

References

Atlan, J. (1999) Possibilités biologiques, impossibilités socials, in H. Allan (ed.) *Le clonage humain*. Paris: Ed. du Seuil, pp. 36–37.

De Miguel Beriain, I. (2008) *La Clonación, diez años después* (Cloning, ten years on). Granada: Comares.

Femenía López, P. J. (1999) *Status jurídico civil del embrión humano, con especial consideración al concebido in vitro*. Madrid: McGraw-Hill.

Jasanoff, S. (2011) Constitutional moments in governing science and technology. *Science and Engineering Ethics*, 17.4: 621–638.

Plomer, A. (2002) Stem cell research in the UK: from parliament to the courts, *Law and the Human Genome Review*, 16: 188 pp.

Romeo Casabona, C. M. (January–July 2006) La cuestión jurídica de la obtención de células troncales embrionarias humanas con fines de investigación biomédica. Consideraciones de política legislativa (The legal question regarding obtaining human embryo stem cells for biomedical research. Legislation policy considerations), *Law and the Human Genome Review*, 24: 75 pp.

Savulescu, J. (April 1999) Should we clone human beings? Cloning as a source of tissue for transplantation, *Journal of Medical Ethics*, 25(2): 87–95.

Star, S. L. and J. R. Griesemer (1989) Institutional ecology, 'translations' and boundary objects: amateurs and professionals in Berkeley's Museum of Vertebrate Zoology, 1907–39. *Social Studies of Science*, 19: 387–420.

Sweden's The Genetic Integrity Act (2006: 351).

7
The Multiplicity of Norms: The Bioethics and Law of Stem Cell Patents

Judit Sándor and Marton Varju

Introduction

Law and ethics present a distinct pathway in the social science analysis of regenerative medicine. They are both normative systems which establish the boundaries of human activities and social interactions following socially recognised value-based considerations. They pursue different social objectives (Adorno, 2009, p. 224), and demonstrate different characteristics. Ethics is more discursive, flexible when determining boundaries in rapidly developing fields, such as biomedicine, and capable of recognising a plurality of non-exclusive viewpoints and value judgements. Law adheres to demands such as certainty, accessibility, clarity, and consistency, follows a mainly binary logic distinguishing between legal and illegal in regulating human activity, and it offers binding normative arrangements enforceable in an attached institutional framework. Its characteristics make law an attractive normative system for the entrenchment and compartmentalisation of boundaries negotiated in ethics.

The principles of contemporary bioethics provide a value-based normative framework for human activities in biosciences and medicine. They are concerned mainly with human intervention with human life and the use of the human body and human biological material. The foremost principle is respect for the dignity and integrity of human beings which addresses practices of objectifying, instrumentalising, commodifying, and commercialising the human body and its parts. The boundaries established in bioethics under the human dignity principle may reflect universal considerations, such as the autonomy of the person, or represent value judgements of different value communities manifested in the diversity of biomedical research regulatory regimes on the national level. In the latter case, any attempt at legal regulation on the regional or global level needs to recognise the multiplicity of local normative arrangements.

In biomedicine, law is responsible for translating into binding rules social and policy expectations of progress and innovation, the demands of commercial stakeholders in the 'bioeconomy', the concerns relating to risk and hazard in human interference with biological matter, and, in particular, the boundaries of human activity in biosciences as indicated in bioethics. These rules may contain prohibitions and threaten the breach of those prohibitions with sanctions, require human activities to be licensed, screened, monitored, and reviewed, indicate how the market may penetrate into scientific activity and how scientific activity may benefit from the existence of a market, offer incentives for scientific progress, and generally provide a clear and predictable framework for actors and stakeholders. A potential shortcoming of law and the legal process is that its coverage may not be comprehensive and may lag behind scientific developments. Law may fail to offer a normative solution for novel scientific and technological developments, or may focus on technologies rendered outdated by new advances in science and technology. Law may also struggle with translating permeable and moving boundaries negotiated extra-legally and with accommodating a plurality of non-exclusive viewpoints on what constitutes good (ethically acceptable) and bad science, or science and non-science.

The normative arrangements of ethics and law are not isolated. Law gradually incorporated the principles of bioethics often prompted by great social controversies, such as abortion, assisted reproduction, cloning, or human embryonic research. It reacted by drawing up rule-based solutions following or dictating social perceptions. The reception in law of bioethical principles was helped by a common, if imperfect, rights-based language associated with human dignity and integrity. The human rights which correspond to the principles of bioethics now form part of the legal regulation of biomedical research, health care and the 'bioeconomy'.

As we saw in Chapter 6, the translation into law of the normative arrangements of bioethics is of central concern in rapidly evolving biosciences and for the profit-oriented stakeholders of the 'bioeconomy'. Discrepancies, ambiguities, and contradictions in the law or the silence of law are a cause of misdirection for actors and policymakers, or enable opportunistic behaviour. The enhanced normative response of the law may, however, prove to be crucial in settling questions at the frontiers of biotechnological development. Law creates 'order by sorting out the complexities of human experience into categories that can be rationally dealt with' and expresses 'binding, collective judgements about the nature of things in the world' (Jasanoff, 2002, p. 895). Law as the gatekeeper of ethically acceptable science produces a final, often quick response entrenching boundaries by selecting between right and wrong (lawful and unlawful) human practices. The credibility and integrity of law depends on how successfully it is able to translate the boundaries deliberated extra-legally in bioethics.

The compartmentalisation of issues and the entrenchment of debates relating to processes of life in the real world in law come with doubts. The gaps and silences of law, the ambiguities of the legal language used, or the quality of legal regulation question whether the law is able to accommodate the complexities of bioethical principles which often require the recognition of multiple, non-exclusive viewpoints, or involve hard choices between competing values. Crucially, in the process of legal interpretation open-ended, ambiguous terms in the law, such as 'research' or 'commercial use', licence the legal forums, agencies and courts, entrusted with the interpretation and application of the law to reassess and possibly redraw the boundaries incorporated into legislation. Following the doctrines or an attractive approach of legal interpretation, these forums may engage in an upward or downward gradation of ethical boundaries and arrange an early closure or a broader playing field in the law. The intervention of such legal forums raises serious legitimacy concerns when their interpretation of the law alters the boundaries recognised in legislation.

In Europe, most controversy has emerged from the parallel existence of the normative principles of bioethics and the rules of intellectual property law, especially patent law. Patent law provides an interface between science and the market, between secluded laboratories and the open public domain. The question of what can be regarded and claimed as intellectual property (i.e., what constitutes patentable subject matter and what inventions may be patentable) is an important boundary for science (Hirsch, 2004, p. 179). The most acute dilemma was whether to extend the socio-legal concepts of invention and intellectual property rights to the human body and human biological material (tissues, cells) and whether the commercial character of the patent system, defined as a commodity system designed to facilitate the commercialisation of human innovative activity, may be at odds with the bioethical principles governing the use of the human body and human biological material.

The boundary was found in the legal distinction between discoveries and inventions (nature and culture) focusing on human contribution capable of transforming products of nature into products of human ingenuity. The other ethical controversy, that the patent system may make certain ethically controversial human scientific practices available in the market, prompted a legislative response, although only in Europe, harmonising the conditions of patentability in national patent laws. The terms used in European legislation in this regard led to contestable interpretative practices before different forums affecting the ethical boundaries expressed therein.

In this chapter, we will examine the distinct journey for regenerative medicine as prompted by the parallel existence of and boundaries between the normative systems of law and ethics. First, we will look into the translation into law of the ethical boundaries of biomedical research, in particular, how law managed to perform this task amidst the complexity of bioethical principles. The recent legal battle as part of the US federal stem cell research funding saga concerning the 2009 National Institutes of Health funding guidelines is used to demonstrate how the legal terms applied to translate bioethical boundaries could lead to the reassessment of those boundaries by courts of law. In the second part, the translation of the relevant bioethical boundaries into intellectual property law relating to the patenting of human biological material will be scrutinised. The patenting of human embryonic stem cells (hESC) and the application of the law to stem cell patents, as demonstrated by the European developments culminating in the judgement of the European Court of Justice (ECJ) in the *Brüstle v. Greenpeace* case, indicate the problems faced by law when incorporating boundaries determined extra-legally and the pressure imposed on those boundaries in the process of legal interpretation.

The ethical boundaries of biomedical research and the law

The emergence of contemporary bioethics is linked to the proliferation of international codes on (bio)medical practices (Nuremberg Code, 1946–1947; WMA International Code of Medical Ethics, 1949; WMA Helsinki Declaration, 1964), international documents connecting bioethics to human rights (Universal Declaration of Human Genome and Human Rights, 1997; UNESCO Universal Declaration on Bioethics and Human Rights, 2005b), and binding international treaties (Oviedo Convention on Human Rights and Biomedicine of the Council of Europe, 1997). These instruments indicate the interest in the gradual legalisation of the field and the related agenda of incorporating into law the principles of bioethics determining the boundaries of human activity in medicine and medical research. The movement towards recognition in the external set of standards of law recognised that law will enable an entrenchment of bioethical principles, and provide enhanced protection for these principles and a coherent and enforceable framework of 'dos and don'ts'.

In bringing the normative arrangements of bioethics and law together, the language of human rights offered an accessible medium of mediation. The apparent success of international and regional human rights instruments, such as the Universal Declaration of Human Rights and the European Convention on Human Rights (ECHR), fuelled an agenda aiming to secure the benefits of international human rights law for bioethics, to connect, in Faunce's terminology to 'subsume' (Faunce, 2005, p. 177), bioethics to the advanced legal, political, and governance framework of the international human rights community. International human rights law is seen as offering a 'global language' (Ashcroft, 2008, p. 49), a global conceptual framework for bioethics enabling it to achieve a status of a universal normative system (Adorno, 2009, p. 227). The United Nations Educational, Scientific, and Cultural Organisation (UNESCO) leading up to the adoption of the Universal Declaration expressed a clear intention to 'unite these two streams' and 'establish the conformity of bioethics with international human rights law' (UNESCO, 2005a, para. 12). Their fusion is facilitated by the fact that both normative systems reflect on the human condition, human suffering and well-being, concepts like human dignity and integrity and personal autonomy providing the necessary overlapping elements. One obstacle of their complete fusion is the theoretical and practical imperfections of their match often characterised by 'mutual incomprehension' (Ashcroft, 2008, p. 48).

The translation of bioethical principles into 'universal' human rights, universal entitlements following from the human condition, is especially difficult considering that socio-cultural, philosophical and religious diversity is a key component of the bioethical discourse. Expressing this diversity on the international and regional level is a major challenge for law, indicated persuasively by the variety of regulatory practices concerning biomedicine in different states reflecting divergent moral viewpoints and compromises in different value communities which need to be accommodated under a single legal framework. Bioethical diversity demands the recognition in legal instruments of that multiplicity by applying 'sufficiently open-ended' terms 'to both capture an agreed general value and allow for (...) interpretation to accommodate a plurality of moral (or legal) perspectives' (Plomer, 2005, p. 15). The linguistic requirements of accommodating diversity in law may lead to serious dilemmas in linguistic expression, in the (legal) interpretation of the terms used, and in determining the boundaries of interpretative discretion of legal forums equipped with jurisdiction to interpret and apply the law. Owing to the open-ended nature, the indeterminacy and ambiguity of the legal text, the particular interpretative methods of law and the particular logic of law in resolving conflicts between competing interests and values will inevitably play a crucial role in completing the fusion of law and bioethics. In such instances, it will remain open to contestation that the result of legal interpretation and the application of the law meet the original intentions of the drafters and the legislator.

The European framework for the protection of fundamental rights and freedoms was especially progressive in the translation of bioethical principles into binding legal arrangements and achieved this having regard to the multiplicity of ethical viewpoints on the national level regarding the boundaries of biomedical research activity. The Oviedo Convention on Human Rights and Biomedicine (see Chapter 6), with a focus on the human rights limitations of biomedical research and therapy, builds on the protection of human dignity and integrity and confirms the primacy of the human being over the interests of science and society (articles 1 and 2). Despite its universalism on the level of general principles, the Oviedo Convention mirrors the diversity of European moral attitudes towards biomedical research. Determining the personal scope of the Convention, in particular of the right to respect for human dignity, is deferred to the national level. The Convention also leaves the question of human embryonic research partially open to domestic discretion by the provision that 'where the law allows research on embryos in vitro, it shall ensure adequate protection of the embryo' (article 18(1)). The more contentious provision of article 18(2) prohibiting the creation of embryos for research purposes in Europe prevented the ratification of the Convention by all Council of Europe States considering it as either excessively liberal or conservative.

Law on the international level prescribing a 'thin' layer of common requirements and recognising the locus of ethical judgements and regulation on the state level is also apparent in how the principles of bioethics found expression under the ECHR. The ECHR, constrained by its subsidiary nature expressed in the so-called 'margin of appreciation doctrine', demonstrated sensitivity to the local ethical appreciation of the boundaries of biomedical research. In the face of value multiplicity, the European Court of Human Rights (ECtHR) would adopt a deferential approach as to the bioethical boundaries recognised under the fundamental right to life and to respect for human dignity in European human rights law.

The jurisprudence of the ECtHR provides a good indication how the translation of the normative requirements of ethics into human rights law can be achieved on the European level. In determining whether a certain moral status of the human embryo could be recognised under the right to life, the lack of a European moral consensus and diversity arising from the increasing regulatory activity in this domain of individual European states played a significant role. In the *Vo v. France* case, the Strasbourg court, echoing an earlier decision that found no evidence that the parties to the ECHR had agreed to a particular solution regarding the right of an unborn child to life (Bruegemann case), declared that the question from what point in the biological existence of humans the right to life begins belongs to the discretion of individual Contracting States (paras 83–85). In the same vein, while the 'potentiality' of human embryos and their 'capacity to become a person' was recognised under the right of respect for human dignity, recognising a full moral status of human embryos was deferred to the jurisdiction of individual Contracting States (*Evans v. the United Kingdom*, paras 54 and 56). The lack of a 'clear common ground among the Member States' prevents the European Court from imposing European boundaries in areas characterised by 'fast moving medical and scientific developments' (*S.H. v Austria*, paras 68. 69, 74).

Under the overarching global or regional legal arrangements, promptly recognised by the European human rights instruments, diversity characterises the legal regulation of biomedical research practices on the local level. There are highly developed and less robust regulatory regimes establishing broader or narrower boundaries for biomedical research, all reflecting the ethical viewpoints of the relevant value community often expressed in the opinions of local ethical advisory bodies. The regulatory perception on the national level of ethical boundaries for biomedical research ranges from prohibitive legal regimes, such as the German based on the Embryo Protection Act prohibiting 'improper uses' of human embryos and the Stem Cell Act prohibiting the importation and utilisation of human embryonic stem cells (hESC), to permissive regimes, such as the British based on the Human Fertilisation and Embryology Act (HFEA) 1990 allowing embryonic research on permitted human embryos subject to licensing, and includes intermediate regimes, such as the French based on the Public Health Code allowing research only on supernumerary in vitro fertilisation (IVF) embryos.

The utilitarian trade-off between the need to protect human embryos and the benefits of stem cell research, a central element for law in determining the boundaries (translating the ethical boundaries) of permissible and prohibited human scientific activity, is based on different regulatory approaches and is fixed on a different measure of gradation in the different legal regimes. The German system uses the legal category of 'exempt stem cells', cells harvested before 1 January 2002 in another state according to the regulations of that state from supernumerary IVF embryos as recognised in the Stem Cell Act, which despite the general prohibitions are available for research compliant with the relevant statutory provisions, especially the Embryo Protection Act, and the major legal principles of the German legal system. A less superficial boundary is used in the UK HFEA, which, recognising that protection should be given to human embryos owing to their potentiality to develop into human beings, applies the '14 day rule' referring to the development of the primitive streak 14 days after fertilisation as the first indication of that potentiality. Research is only permitted on embryos in the 14 days after fertilisation using supernumerary IVF embryos, IVF research embryos, and embryos created for research purposes by somatic cell nuclear transfer (SCNT). The French legal approach focuses on the sources of human embryos following the opinions of the National Ethics Committee on the ethical acceptability of creating and using embryos for research purposes (*Avis* 8, 52, 54). It allows only the use of supernumerary IVF embryos from a parental project for research, in addition to aborted dead embryos and foetuses from which cells and tissues may be collected (article L1241-5, Public Health Code).

The ethical foundations of these legal boundaries are often transparent. The law in France appears to follow the ethical boundaries determined by the expert National Ethics Committee. The boundaries in place in the Public Health Code received further legal and constitutional confirmation in a decision of the French Constitutional Council (Decision 94/343/344). The rules of the UK HFEA are declared to reflect social consensus developed in decades of social deliberation and discussion concerning the moral status of human embryos and the ethical boundaries of biomedical research activity (Stem Cell Research and Regulations; Human Fertilisation and Embryology Regulations). The exemption of the German Stem Cell Act seems to lack similarly robust ethical foundations as demonstrated by the deep divisions within the German Ethics Council on this matter (see Nationaler Ethikrat, 2001).

In determining the boundaries of biomedical research, the legal regimes of Japan, South Korea, India, and China demonstrate remarkable similarities with Western regimes, especially the United Kingdom (Medical Research Council -CURE Report, 2009). They regulate the acceptable sources of human embryos for research, and rely on legal distinctions in legitimising research activity, such as the '14 day rule' or a similar rule referring to the development of the primitive streak (China Guiding Principles; India Guidelines; South Korea Bioethics and Safety Act; Japan Human Cloning Techniques Act). Concerning the underlying ethical viewpoints, we have limited information available. Internationalisation (international 'principlism') and borrowing from foreign jurisdictions may account for the legal developments as indicated in reports on the impact of international (universal) principles on the Chinese regulatory approach (Döring, 2003, pp. 233, 236; McMahon and Thorsteinsdóttir, 2010, p. 290). Striking a balance between the protection of values reflected in the recognised bioethical principles and promoting development in biomedical sciences is regarded as the core paradigm of the Chinese regulatory regime (China Guiding Principles; China Ethical Review Regulations; UK Stem Cell Initiative country report).

The practical approach to regulating bioethical boundaries in South Korea is explained by the dominant social doctrine of scientism seeing science as source of growth and development for the nation (Harmon, 2008, pp. 268, 281) and by considering advances in biosciences and embryology as a social and economic triumph. A 2010 judgement of the South Korean Constitutional Court provides a clearer indication of the background of legal boundaries (Case 2005/346) holding that early-stage human embryos (between conception and 14 days after conception) are not subjects of human rights. It argued that early embryos may not be conceived as human beings as they only represent the beginning of the process of developing human life, the development of the primitive streak being the point from which law may treat them as independent human beings. The Japanese regulatory approach is based on a compromise between acknowledging that the human embryo represents the beginning of human life and that it serves as a crucial resource for socially valuable research (Japan Guidelines). The law focuses on prudent scientific activity and the strict monitoring of that activity (Sleeboom-Faulkner, 2009), and places lesser emphasis on value of human embryos: human embryos are not conceived as persons or subjects with rights, only as entities that deserve respect in their treatment (Matsuda, 2007).

The recent legal developments in the US debate on the federal funding of stem cell research provide a good indication of the vulnerability of bioethical boundaries as translated into law. The intervention of federal courts brought to light how incomplete the legal terms used to delineate bioethical boundaries may be and by way of legal interpretation before the legal forums with jurisdiction to interpret and apply the law how those boundaries could be subject of modification or gradation on a scale of moral permissiveness. In the United States, human embryonic research is not regulated directly on the federal level; if regulated, it is regulated differently in separate federal states. Apart from the federal guidelines on research involving human subjects (Guidelines), only foetal research is subject to federal regulation under section 289 G of the US Code introduced in the wake of the landmark abortion decision by the US Supreme Court in *Roe v. Wade* (410 US 113 (1973)). The only discernible element on federal level is the regulation of federal funding for public research through the responsible agency, the National Institutes for Health (NIH), not affecting privately funded research.

The ethical boundaries of policy and regulation on the federal level were subject to substantive changes in the past decades. The ethical advisory committees appointed by successive administrations produced contradicting viewpoints on the moral acceptability of hESC research. President Clinton's National Bioethics Advisory Commission agreed in its 1999 Report that supernumerary IVF embryos should be allowed to be destroyed in the process of generating 'stem cells for bona fide research' (NBAC, 1999). The Bush administration's President's Council on Bioethics redrew the boundaries in its 2005 Report by contending that the protection of human life from the earliest stages of development, including the human embryo, is a widely accepted ethical norm which precludes the seeking of therapies by means of destroying human embryos (PCB, 2005).

These alterations were adequately mirrored in the funding policy on hESC research with alternating periods of federal funding being available and being frozen. The Clinton administration's liberal policy (HERP Report, 1994; Parens, 2001, p. 37) was reversed by President Bush Jr. excluding from federal funding the derivation of further hESC lines from embryos (*Address to the Nation*; EO No. 13435). In Executive Order No. 13435, examining whether in the process of cell line derivation human embryos were destroyed, discarded or subjected, to harm, hESC derivation was seen as violating the principle of non-commodification and in breach of the premise that human embryos are 'members of the human species'. The boundaries were shifted again by the Obama administration which in Executive Order 13505 removed the federal funding moratorium claiming that the interest of progress in hESC research for the purposes of enhancing human biomedical knowledge and creating new therapies prevails over competing socio-ethical considerations. It indicated that research will be eligible for federal funding only when it is responsible and scientifically worthy and permitted by law, the latter point leading to further legal controversies.

The only concrete legislative measure in the federal funding debate is the 1996 Dickey–Wicker amendment, a rider attached to the Balanced Budget Downpayment Act 1996, implementing a ban on spending federal monies on stem cell research. It prohibited the use of federal funding for the creation of human embryos for research purposes and research in which human embryos are destroyed, discarded, or knowingly subjected to risk of injury or death greater than that allowed for research on foetuses under federal regulation. The measure was a serious constraint on the liberalisation of the federal funding policy under the Obama administration. The 2009 NIH *Guidelines for Human Stem Cell Research*, establishing the rules for federal funding after Executive Order 13505, were developed having regard to the Dickey–Wicker amendment, and provide that NIH federal funding cannot be used for the derivation of stem cells from human embryos, and research using hES cells derived from other sources, including SCNT, parthenogenesis, and/or IVF embryos created for research purposes, is not eligible for NIH funding.

Pushing for a more restrictive interpretation of federal funding policy under the Dickey–Wicker amendment, in 2010 an injunction was served by a US District Court against the spending of federal funds under the 2009 NIH Guidelines (*Sherley v. Sebelius*, 704 F. Supp. 2d 63, 70 (D.D.C. 2010)). The 'Lambert injunction' contended that making federal funding available under the Guidelines was in breach of the Dickey–Wicker amendment which in the court's view explicitly prohibited the use of federal funding for research in which a human embryo or embryos are destroyed. The availability of funding for existing hES cells was also found illegal as in the court's interpretation the Dickey–Wicker amendment's intent, lacking an express limitation to the contrary, covered all research activity associated with stem cell derivation.

The 'Lambert injunction' is a clear indication that the boundaries entrenched in legislation could be shifted before legal forums with jurisdiction to interpret the law contradicting policy developed under those boundaries. It relied on the terms used in the Dickey–Wicker amendment to move the legal boundary upwards by interpreting hESC research as necessarily depending upon 'the destruction of the human embryo' and the derivation of hESC from an embryo being 'an integral step' in conducting hESC research. The judge's perception of the linguistic boundary of 'research', the term used in the Dickey–Wicker amendment, had a direct impact on the bioethical boundaries prescribed for the federal funding of stem cell research.

In response to the injunction, the NIH suspended the financing of research on the available hESC lines. However, a few weeks later the 'Lambert injunction' was ordered to be stayed in appeal (*Sherley v. Sebelius*, Court of Appeal, No. 10-5287, 9/9/2010) and the NIH lifted the suspension on funding. The final appeal decision found an interpretation of the law different from the District Court and allowed federal research funding while the lower court reexamines the case (*Sherley v. Sebelius*, Court of Appeal, No. 10-5287, 29/4/2011).

The tenor of the appeal court decision was that having regard to the exact terms used in legislation it was 'entirely reasonable' for the NIH to interpret the Dickey–Wicker amendment as indicated in the 2009 Guidelines. Following the text of the legislation, the boundary between eligible/ineligible (ethical/unethical) was found by distinguishing between research on the existing stem cell lines derived without federal funding and research for the derivation of further stem cell lines seeking federal funding. There is a clear split between what is considered in law as ethical when research is funded by private money and when funded from federal monies, and between the morality of research on existing stem cell lines and of the actual process of deriving stem cell lines. All turned on the reading of statutory terms by the court and on the application of the relevant legal doctrines relating to statutory interpretation.

The reexamination of the case by the District Court (*Sherley v. Sebelius*, Memorandum Opinion (27/7/2011)) endorsed the interpretation developed by the Court of Appeal. The statement that the terms used in the Dickey–Wicker amendment, such as 'research', were ambiguous 'as a matter of law' was found binding in the lower court which accepted that the NIH's Guidelines 'were based on a permissible construction of the statute'. In the court's view, following the appeal judgement, the interpretation of the immediate terms surrounding the prohibition of the Dickey–Wicker amendment allow the conclusion that hESC research is not 'research in which a human embryo or embryos are destroyed' and 'research in which a human embryo or embryos are . . . subjected to risk of injury or death'.

The legal developments in the US federal funding debate is an outstanding example how the legal process may be able to impose its own considerations on the translation of bioethical boundaries into law. Legal boundary work could become a matter for legal interpretation, a 'linguistic jujitsu', as stated by Judge LeCraft Henderson in her dissenting opinion in the appeal judgement, over the terms of legislation in order to determine (recalibrate) the original legislative intent. The legal interpretation of statutory terms, such as 'research', by different legal forums could lead to an upward gradation of bioethical boundaries without paying attention to the underlying bioethical debate or consensus. In the *Sherley v. Sebelius* decisions, the linguistic difference between 'research in which', 'research from which', and 'research as a result of which' (human embryos are destroyed or harmed) presented the licence for the courts to reassess the ethical boundary laid down in federal legislation. With so much depending on the subsequent interpretation and application of the law, it is open to debate whether the legislative intention to fuse the parallel normative empires of law and ethics could be achieved. The law's preoccupation with its intrinsic normative arrangements could lead to miscommunication between the two domains.

Bioethical boundaries and the law of stem cell patents

A rather similar problem concerning the interpretation of legal terms delimiting the patentability of human biological material, hES cells in particular, arose in the specialised legal domain of patent law. Patent law is a traditionally ethically 'sterile' area of law, its main function being the channelling of innovation to the market. Patents serve as incentives for innovation; they are an interventionist instrument from the state to foster progress. The patent system ensures that inventions (knowledge) are brought into the public domain for the benefit of society and inventors benefit from the commercial opportunities presented by the limited monopoly granted by patent law. Patents offer a reward for the socially valuable activity of generating an inexhaustible resource, knowledge. Promoting innovation as a public policy and looking after the private economic interests of the inventor and other economic stakeholders (investors, commercial developers) are at the heart of patent law.

The patent system could be defined as a commodity system which invests knowledge and innovation with a commercial value. It treats inventions, including the materials constituting an invention, as commodities. Within the patent system, things are treated as objects with utility which are fungible and available for commercial exchange. As its main commercial function, the patent system by introducing innovation to the market enables their future industrial and commercial exploitation.

With the emergence of 'bio-objects' (see Vermeulen et al., 2011), (human) biological material used as base material for biomedical research and therapies, the normative arrangements of patent law were put on a collision course with the normative system of bioethics. The relevance of bioethics in the patenting process was brought to light when the law redrew the boundaries of what may constitute an invention, shifting the dividing line between nature and culture, by establishing that biological material could be brought under the scope of the patent system as patentable subject matter provided there is a significant element of human intervention. This strategic change in patent law, which enabled a new range of emerging sciences to access the benefits of the patent system and the patent system to engulf the material wealth promised by these new domains, left the patent system exposed to contestation and counter-claims emerging from the principles of bioethics. Treating biological material isolated from the human body as useful objects of commercial value is a development which raises concerns under the core principle of respect for human dignity.

The exposure of patent law to the demands of bioethics needs reassuring responses. Patent law cannot avoid expressing the trade-off enabling the patenting of human biological material between observing the bioethical principles relating to human life and the human body and recognising the social and individual benefits of biomedical progress. Also, in order to enable the opening of the commodity system known as the patent system to human biological material, often associated with ethically controversial human interventions with human life and the human body, patent law needs ensure that it complies with the relevant bioethical principles derived from human dignity, such as non-instrumentalisation, non-commodification, and non-commercialisation. In global, regional, and local patent law the responses to these bioethical challenges were somewhat patchy, European patent law providing the only exception, which leaves a considerable gap between the normative universes of law and ethics.

Despite the pressures of its broader regulatory environment, US patent law remained faithful to its purposes stated in the US constitution, especially in section 8, that of securing the successful commercialisation of inventions. It was the first patent system to create a new boundary for innovative activity by recognising the patentability of biological material, material available in nature, on the condition that its 'invention' involved recognisable human involvement. In the judgement of the US Supreme Court on the *Diamond v. Chakrabarty* (447 US 303 (1980)) a majority of five justices recognised the legal distinction between products of nature and of human ingenuity and found that the human intervention of isolating living matter is sufficient to distinguish between mere discoveries and inventions of biological material. The real implications of the judgement were revealed some time later when the patent claims for the Harvard Oncomouse and the Wisconsin Alumni Research Foundation (WARF) stem cells were made. The new boundary, however, left little space for bioethical considerations in US patent law (Jasanoff, 2002, p. 95; Filliben, 2008–2009, p. 243), which as an exception from global trends does not contain a clause making exceptions from patentability on *ordre public* and public morality grounds. Without Congress legislating on this matter, US patent law will not address the ethical dilemmas arising out of *Diamond v. Chakrabarty*. Bioethical constraints should be expressed in the separate set of norms of biomedical research regulation which, as we saw earlier, on the federal level suffers from gaps and from its own boundary problems.

Patent laws in other relevant jurisdictions were hardly more receptive towards the normative requirements of bioethics. While patent laws in the biotechnology powerhouses of Asia recognise that the patent system may be closed for inventions on public morality grounds (article 5 of Patent Act in China, article 3b of the Patent Act of 1970 in India, article 32 of the South Korean Patent Act, and article 32 of the Patent Act of 1959 in Japan), there is not much indication that the public morality clause would incorporate the boundaries established in bioethics and that it would be applied to exclude from patentability inventions consisting of or containing human biological material. As an exception, the Indian patent office's draft manual recognised 'method of cloning' as an invention which would 'violate the well accepted and settled social, cultural, legal norms of morality' (Point 4.3).

Outside of Europe, the infiltration of bioethical considerations into patent law is perhaps most visible in Canada. In relation to the question of patentability, the Canadian Intellectual Property Office's (CIPO) *Stem Cell Notice* distinguishes between higher and lower life forms, and holds that fertilised eggs, embryos, and totipotent stem cells are higher life forms and unpatentable, and pluripotent and multipotent stem cells, which do not have the potential to develop into a higher life form, are patentable subject matter. The Canadian position was suggested to be influenced by European practice and the UK Intellectual Property Office's 2007 notice on stem cell patents (Hagen, 2008, p. 516).

The most comprehensive incorporation of ethical principles relating to the treatment of the human body and human biological material as objects of utility and of commercial relevance by the patent system is provided in European patent law, harmonised by European Union (EU) Directive 98/44/EC. The Directive was to respond to a number of different demands, such as those of EU economic and innovation policy envisioning a dynamic and competitive European biotechnology sector, those of EU market regulation aiming at the reduction of regulatory differences in national patent laws, and the social demands for delineating the boundaries of patenting the human body and biological material harvested from it. The Directive guaranteed a priority position for the relevant bioethical principles and set out to achieve its regulatory and policy objectives having regard to those principles.

The result was a complex regulatory blueprint for national patent laws, and eventually for the European Patent Convention establishing the European Patent Organisation (EPO), incorporating ethically informed distinctions regarding what may constitute patentable subject matter and the patentability of what inventions should be rejected on account of their commercial exploitation being contrary to the requirements of public morality. Articles 3 and 5 of the Directive, by introducing the conceptual distinction between inventions and discoveries, hold that (human) biological material including the parts of the human body are patentable subject matter provided they were subject to human intervention by isolation or production by means of a technical process. According to the preamble of the Directive, the boundaries of what may constitute patentable subject matter were redrawn in European legislation having regard to the significant social and economic utility of (human) biological material (Recitals 17 and 18). By way of ensuring that the ethical principle of respect for human dignity and integrity is observed, article 5 of the Directive excludes the human body at any stage of its formation and development from the scope of European patent law preventing its commodification in the patent system.

The commodification of 'bio-objects' in the patent system enabled by article 3 and 5 of the Directive is based on a utilitarian trade-off allowed by the principles of European bioethics. The Directive was created under the assumption that there is European consensus supporting the redrawing of boundaries of what may constitute inventions under patent law by accommodating the scientific developments of biotechnology. The

ECJ saw these provisions as compatible with the requirement of respect for human dignity ensuring that *only the result of inventive, scientific or technical work are patentable and biological information and material existing in their natural state in the human body are only patentable 'where necessary for the achievement and exploitation of a particular industrial application'* (para 75, C-377/98).

Further bioethical principles found expression in article 6 of the Directive which excludes from patentability inventions the commercial exploitation of which would contravene public morality. Paragraph 2 identifies specific inventions the availability of which through the patent system in the market for commercial exploitation should be excluded. It mentions inventions, such as processes of cloning of human beings, processes for modifying the germ line genetic identity of human beings, and uses of human embryos for industrial or commercial purposes. These specific examples reflect a European consensus on the ethical unacceptability of these scientific interventions with human life. The boundaries under the general public morality clause are less certain as it should be regarded as capable of accommodating the multiplicity of ethical viewpoints among European states as expressed in European law and in the national regulation of biomedical research. This is clearly acknowledged in Recital 39 of the Directive stating that '*ordre public* and morality correspond in particular to ethical or moral principles recognised in a Member State' and was confirmed in the ECJ interpretation of article 6 (paras 37–39, C-377/98).

The marriage of the normative provisions of bioethics and the norms of European patent law under article 6 of the Directive is not the most convenient. The difficulties lie not with the silences and gaps in the legal regulation but with the open texture of the legal terms. The boundaries established in Article 6 are open to upwards and downwards gradation, their linguistic expression exposing them to the interpretative discretion of legal forums entrusted with the interpretation and application of law. The crucial 'industrial or commercial use of embryos clause' fails to determine with precision what element of commercial or industrial significance and which segment of the innovative process would prompt the exclusion of patentability under that clause enabling in the application of the Directive the shifting of boundaries on a vertical scale of ethical permissiveness and prohibitiveness.

In the short European history of patenting hESCs, the interpretative efforts relating the ethical boundary recognised in the 'industrial or commercial use clause' under article 6(2)c of the Directive presented the most controversial chapter. It enabled the accommodation within the realm of patent law the broader bioethical concerns associated with hESC research, in particular, that at current state of art the harvesting of stem cells from the human embryo necessitates its destruction. It also allowed recognising the act of submitting a patent claim as the commercial element capable of triggering the application of article 6(2)c to deny the patentability of inventions, instead of focusing on the question whether the invention and its availability in the market through the patent system would lead to the exploitation of human embryos on an industrial scale and/or pursuing commercial aims.

The application by different European forums of article 6 of the Directive to hESC patents highlights the vulnerability of the legal entrenchment of bioethical norms to the problems associated with legal language and interpretation and questions whether an adequate integration of the normative domains of ethics and law was achieved in the Directive. The patentability of inventions consisting of or containing stem cells was decided on the basis of the 'industrial or commercial use clause' under article 6(2)c, which, in the interpretation of the organs of the EPO, should attract a broad interpretation and excludes from patentability not only uses of human embryos but also hES cells derived from human embryos leading to the destruction of those embryos (see the *Edinburgh Patent Decision* [EPO, 2002]; and the *WARF Decision* [EPO, 2008]). The bioethical boundary expressed in article 6(2)c attracted an interpretation that since the performing of the invention with an intention to patent that invention constitutes the commercial element which invokes the application of the 'industrial or commercial use clause', any involvement of human embryos in the preparation of the invention, which may involve the destruction of those embryos, will be deemed as an 'integral and essential part of the industrial or commercial exploitation of the claimed invention' (*WARF Decision*, para. 25).

The final authority on how the clause concerning the industrial or commercial use of human embryos should be interpreted and applied in patent law was produced in the *Brüstle v. Greenpeace* case by the ECJ. It contended that under article 6(2)c the patentability of inventions consisting of or containing hES cells must be excluded on the ground that applying for a patent with the intention to exploit the commercial rights derived from patents would render the invention, the relevant innovative human activity, to be of a commercial or industrial nature. This would arise irrespective of whether the human activity in question, basic biomedical research, would have an essentially non-commercial nature. This is a reading similar to that in the *WARF Decision* focusing on the grant of a patent, on access to the protection provided by patent law, which was found sufficient to render the use of human embryos for the purposes of scientific research activity representing an industrial or commercial use of those embryos (*Brüstle v. Greenpeace*, para 41). Patenting was seen as 'connected with acts of an industrial or commercial nature', and while scientific research was distinguished from industrial and commercial activities, it was argued that 'the use of human embryos for the purposes of research which constitutes the subject-matter of a patent application cannot be separated from the patent itself and the rights attaching to it' (*Brüstle v. Greenpeace*, paras 42, 43).

By declaring that preparing an invention with the intention of patenting that invention provides the commercial element required under the 'industrial or commercial use clause', EPO practice and the ruling of the EU Court of Justice drew a cautious boundary for human innovative activity in European patent law. Arguably, the boundary found through a broad interpretation that article 6(2)c may represent an upward gradation of the boundary established by the European legislator which, if read more narrowly, should exclude from patentability inventions only which constitute an industrial or commercial use of human embryos. Making a patent application may not be regarded as conclusive evidence that the research activity was conducted pursuing commercial interests or that the invention, which involves the destruction of human embryos, would actually be used for industrial or commercial purposes.

A more worrying aspect of the interpretation of article 6(2)c by European forums is the prominent position granted to the argument that human embryos are destroyed in the process of harvesting stem cells. The *WARF Decision* appears to suggest that public morality is violated by 'performing the invention, which includes the step (of destroying a human embryo)' (paras 27 and 29). The approach of the subsequent EPO 'California Stem Cell Decision' is more transparent holding that the destruction of human embryos in the derivation hES cells is sufficient to have the patentability of inventions consisting of or containing hES cells excluded under the 'industrial or commercial use clause' (para 7). Recognising by interpretation in the domain of (patent) law a European bioethical principle condemning the destruction of human embryos in stem cell research would indicate a significant upwards shifting of bioethical boundaries. The ethical background of European patent law's 'human embryo destruction principle' is rather ambiguous and would indicate law assuming a more active role than the mere recording of the principles developed in European bioethics.

The ECJ clearly perceived its licence much more limited than to recognise an 'embryo destruction principle' in its reasoning. There is no indication in the *Brüstle* case that the destruction of human embryos in the process of harvesting hES cells played a role in declining the patentability of inventions consisting of or containing hES cells. The only sign that the 'industrial or commercial use clause' would encompass the moral rejection of destroying human embryos for research purposes follows from a statement when responding to the question concerning the gap between the technical teaching as provided in the patent claim and the actual invention. The Court held that

> an invention must be regarded as unpatentable, even if the claims of the patent do not concern the use of human embryos, where the implementation of the invention requires the destruction of human embryos. In that case too, the view must be taken that there is use of human embryos within the meaning of Article 6(2)(c) of the Directive. (para 49)

The recognition within article 6(2)c that an 'embryo destruction principle' would put considerable strain on the relationship between law and ethics in European 'biopatenting'. Such interpretation of the Directive assumes a moral consensus on this matter among European states which would conflict with the provisions of the Oviedo Convention, the jurisprudence under the ECHR, and the diversity of domestic biomedical research regulatory regimes. Overlooking the limits on introducing bioethical boundaries on the international or regional level would significantly undermine the legitimacy of European patent law. Nevertheless, recognising a moral prohibition of destroying human embryos for research purposes is not excluded under European patent law. The solution is offered by the general public morality clause of the Directive under which the multiplicity of moral viewpoints in European states could be acknowledged and the assessment of patentability with regard to a local 'embryo destruction principle' can be deferred to the competence of national forums.

Conclusion

The translation of bioethical boundaries into European patent law presented a controversial chapter in the global push towards the fusion of bioethics and law. As in case of the US federal funding debate, the cause of controversy was the linguistic expression of bioethical boundaries in law and the subsequent interpretation by different legal forums of the terms used in legislation. The restrictive interpretative approach followed by the EPO Organs and the EU Court of Justice generated a cautious interpretation of the Directive which by every indication led to a gradation, if not the redrawing of the bioethical boundaries recognised in legislation. Communication between the parallel normative regimes of ethics and law does not conclude with incorporating bioethical principles into legislation or international documents. Instead, as we have shown, it will inevitably involve the process of interpreting and applying those principles in law. How this is itself a reflection of wider dynamics at work that shape the regenerative medicine field is explored in the next chapter. There Brian Salter explores the ways in which regional competition at the international level (between, for example, the EU, the United States, and China) leads to ongoing innovation in regimes of governance designed to enable development of the field, but at the same time limits in the ways in which such governance can actually be reconfigured to do so.

References

Adorno, Roberto (2009) Human dignity and human rights as common grounds for global bioethics, *Journal of Medicine and Philosophy*, 34(3): 223–240.

Ashcroft, Richard (2008) The troubled relationship between bioethics and human rights, in Michael Freedman (ed) *Law and Bioethics*. Oxford: Oxford University Press, pp. 32–52.

Döring, Ole (March 2003) China's struggle for practical regulations in medical ethics, *Nature Reviews Genetics*, 4(3): 233–239.

Faunce, Thomas A. (March 2005) Will international human rights subsume medical ethics? Intersections in the UNESCO Universal Bioethics Declaration, *Journal of Medical Ethics*, 31(3): 173–178.

Filliben, Vincent J. (2008–2009) Patent law and regenerative medicine: a consideration of the current law and public policy concerns regarding upstream patents, *Wake Forest Intellectual Property Law Journal*, 9(3): 238–258.

Hagen, Gregory R. (2008) Potency, patenting and preformation: the patentability of totipotent cells in Canada, *SCRIPTed*, 5(3): 515–552.

Harmon, Shawn H. E. and Na-Kyoung Kim (2008) A tale of two standards: drift and inertia in modern Korean medical law, *SCRIPTed*, 5(2): 267–293.

Hirsch, Eric (2004) Boundaries of creation: the work of credibility in science and ceremony, in Eric Hirsch and Marilyn Strathern (eds) *Transactions and Creations: Property Debates and the Stimulus of Melanesia*. Oxford: Berghahn Books, pp. 176–192.

Jasanoff, Sheila (June 2002) The life sciences and the rule of law, *Journal of Molecular Biology*, 319(4): 891–899.

Matsuda, Jun (2007) The Regulations for Research on Human Embryos in Japan and Germany. Manuscript available at http://www.hss.shizuoka.ac.jp/shakai/ningen/staffs/matsuda/20070622.pdf

McMahon, Dominique S. and Halla Thorsteinsdóttir (2010) Lost in translation: China's struggle to develop appropriate stem cell regulations, *SCRIPTed*, 7(2): 283–294.

Medical Research Council (2009) China–UK research ethics (MRC-CURE) report. Available at http://www.mrc.ac.uk/Utilities/Documentrecord/index. htm?d= MRC006303

Parens, Erik (2001) On the ethics and politics of embryonic stem cell research, in Suzanne Holland, Karen Lebacqz, and Laurie Zoloth (eds) *The Human Embryonic Stem Cell Debate: Science, Ethics, and Public Policy*. London: MIT Press, pp. 37–50.

The President's Council on Bioethics (2005) *Alternative Sources of Human Pluripotent Stem Cells: A White Paper*. Washington, DC: The President's Council on Bioethics.

Plomer, Aurora (2005) *The Law and Ethics of Medical Research: International Bioethics and Human Rights*. London: Cavendish Publishing.

Sleeboom-Faulkner, Margaret (2009) *Human Embryonic Stem Cell Research (hESR) in East Asia: An Institutional Approach to Bioethical Reorientation*, Full Research Report, ESRC End of Award Report, RES-350-27-0002 (Swindon: Economic and Social Research Council).

Vermeulen, N., T. Sakari and A. Webster (eds) (2011) *Bio-Objects: Life in the 21st Century*, Farnham: Ashgate Publishing.

World Medical Association (1949) *International Code of Medical Ethics*. Available at http://www.wma.net/en/30publications/10policies/c8/

World Medical Association (1964) *Declaration of Helsinki – Ethical Principles for Medical Research Involving Human Subjects*. Available at www.wma.net/e/policy/ b3.htm

Legal Cases and Documents

Act on Regulation of Human Cloning Techniques of 2000, Japan, available at http://www.japaneselawtranslation.go.jp/law/detail/?re= 02&dn= 1&x= 0&y= 0&co= 1&yo= &gn= &sy= &ht= &no= &bu= &ta= &ky= cloning&page= 1

Bioethics and Safety Act of 2004 (South Korea), available at http://eng.bprc.re.kr/ gz06.htm?number= 8

Canadian Intellectual Property Office (2006) Practice Regarding Fertilized Eggs, Stem Cells, Organs and Tissues. Available at http://www.cipo.ic.gc.ca/epic/site/ cipointernet-internetopic.nsf/en/wr002953.html

China: Ethical Guiding Principles on Human Embryonic Stem Cell Research. Available at www.qmlc.com.cn/edit/UploadFile/info/2009430113029216. doc

The China 2007 (11 January) new Regulation on Ethical Review of Biomedical Research involving Human Subjects, available at http://www.mrc.ac.uk/ Utilities/Documentrecord/index.htm?d= MRC006303

China: UK Stem Cell Initiative country report, available at http://www. advisorybodies.doh.gov.uk/uksci/global/china.htm

Code de la Santé Publique. Available at http://www.legifrance.gouv.fr/affichCode. do?cidTexte= LEGITEXT000006072665&dateTexte= 20101017

Comité Consultatif National d'Éthique, CCNE (1986) *Avis 8*. Available at http:// www.ccne-ethique.fr/avis.php

Comité Consultatif National d'Éthique, CCNE (1997) *Avis 54*. Available at http://www.ccne-ethique.fr/avis.php

Conseil Constitutionnel (1994), Decision 94/343/344 DC. Available at http://www.conseil-constitutionnel.fr/conseil-constitutionnel/francais/les-decisions/acces-par-date/decisions-depuis-1959/1994/94-343/344-dc/decision-n-94-343-344-dc-du-27-juillet-1994.10566.html

Council of Europe (1997) Convention for the Protection of Human Rights and Dignity of the Human Being with Regard to the Application of Biology and Medicine: Convention on Human Rights and Biomedicine. Available at http://conventions.coe.int/Treaty/en/Treaties/Html/164.htm

Diamond v. Chakrabarty, 447 US 303 (1980).

Dickey-Wicker Amendment, Pub. L. No. 104-99, § 128, 110 Stat. 26, 34 (1996)

Directive 98/44/Ec of the European Parliament and of the Council Of 6 July 1998 On The Legal Protection of Biotechnological Inventions. Available at eur-lex.europa.eu/LexUriServ/LexUriServ.do?uri=OJ:L:1998:213:0013:0021:EN:PDF

ECommHR: *Brueggemann and Scheuten v. Germany* (1981) 3 EHRR 244.

Embryo Protection Act, Germany. Available at http://www.bmj.bund.de/files/-/1147/ESchG englisch.pdf

Ethical Guidelines for Stem Cell Research (India). Available at http://www.icmr.nic.in/stem_cell/stem_cell_guidelines.pdf

European Court of Human Rights (2000) *S.H. and Others v. Austria*, App. 57813/00, nyr

European Court of Human Rights (2004) *Vo v. France*, App. 53924/00, 08 July 2004, ECHR 2004-VIII.

European Court of Human Rights (2007) *Evans v. United Kingdom*, App. 6339/05, 10 April 2007, nyr.

European Court of Justice (1998) *The Netherlands v. Council and Parliament*, Case C-377/98 [2001] ECR I-7079.

European Court of Justice (2010) *Brüstle v. Greenpeace*, Case C-34/10, nyr.

European Patent Office (2008) Wisconsin Alumni Research Foundation (WARF) Decision, EPO Enlarged Board of Appeal Decision G 2/06 of 25 November 2008, *Official Journal of the European Patent Office*, May 2009, 306.

European Patent Office (2009) *California Stem Cell Decision*, EPO Boards of Appeal Decision T 522/04 of 28 May 2009, nyr.

European Patent Office (2002) *Edinburgh Patent Decision*, the case concerning European Patent No. EP0695351, nyr.

European Patent Office: Interview with Dr Ingrid Schneider. Available at http://documents.epo.org/projects/babylon/eponet.nsf/0/F172DE5BB2B9B15BC12572DC0031A3CB/$File/Interview_Schneider.pdf

Human Ethics Research Panel (1994) *Report of the Human Embryo Research Panel*. Washington, DC: National Institutes of Health.

Human Fertilisation and Embryology (Research Purposes), (United Kingdom) Regulations 2001/188. Available at http://www.legislation.gov.uk/uksi/2001/188/contents/made

Human Fertilisation and Embryology Act (HFEA) of 1990 (United Kingdom). Available at http://www.dh.gov.uk/en/Publicationsandstatistics/Publications/PublicationsLegislation/DH_080205

Ministry of Education, Culture, Sports, Science, and Technology of Japan (2009) *Guidelines for Derivation and Utilization of Human Embryonic Stem Cells*. Available at http://www.lifescience.mext.go.jp/files/pdf/32_90.pdf

National Bioethics Advisory Commission (1999) *Ethical Issues in Human Stem Cell Research*. Available at http://bioethics.georgetown.edu/pcbe/reports/past_commissions/nbac_stemcell1.pdf

National Conference of State Legislatures (NCSL) *Embryonic and Fetal Research Laws*. Available at http://www.ncsl.org/issues-research/health/embryonic-and-fetal-research-laws.aspx

National Institutes of Health (2009) *Stem Guidelines for Human Stem Cell Research*. Available at http://stemcells.nih.gov/policy/2009guidelines.htm

National Institutes of Health (2010a) *Director's Statement on Lambert*. Available at http://www.nih.gov/about/director/08262010statement_stemcellinjunction.htm

National Institutes of Health (2010b) *Stem Cell Statement*. Available at http://www.nih.gov/news/09102010_stemcell_statement.htm

Nationaler Ethikrat (2001) *Opinion on the Import of Human Embryonic Stem Cells*. Available at http://www.ethikrat.org/_english/publications/stem_cells/Opinion_Import-HESC.pdf

Nuremberg Code (1946–1947). Available at http://ohsr.od.nih.gov/guidelines/nuremberg.html

Patent Act of 1959 (Japan). Available at http://www.japaneselawtranslation.go.jp/law/detail/?ft=1&re=02&dn=1&x=0&y=0&co=01&ky=patent&page=17

Patent Act of 1970 (India). Available at http://www.patentoffice.nic.in/ipr/patent/patents.htm

Patent Act of China. Available at http://www.sipo.gov.cn/sipo_English/laws/lawsregulations/200804/t20080416_380327.html

Patent Act of South Korea. Available at http://park.org/Korea/Pavilions/PublicPavilions/Government/kipo/law/patent/epat.html

Patent Office of India (2008) *Draft Manual of Patent Practice and Procedure*. Available at http://ipindia.nic.in/ipr/patent/DraftPatent_Manual_2008.pdf

Roe v. Wade (410 U.S. 113 (1973)).

Sherley v. Sebelius, 704 F. Supp. 2d 63, 70 (D.D.C. 2010).

Sherley v. Sebelius, United States Court of Appeals for the District of Columbia Court, No. 10-5287, 9 September 2010.

Sherley v. Sebelius, United States Court of Appeals for the District of Columbia Court, No. 10-5287, No. 10-5287, 2011 WL United States Court of Appeals for the District of Columbia Court, No. 10-5287, 29 April 2011.

Sherley v. Sebelius, Memorandum Opinion (D.D.C. 27 July 2011).

South Korea Case 2005/346 (27/05/2010). Available from interview with Prof. Na-Kyoung Kim, Singshin University, REMEDiE, December 2010.

Stem Cell Act, Germany. Available at http://www.bmj.bund.de/files/-/1146/Stammzellgesetz englisch.pdf

Stem Cell Research and Regulations under the Human Fertilisation and Embryology Act of 1990 (United Kingdom). Available at http://www.parliament.uk/documents/commons/lib/research/rp2000/rp00-093.pdf

The President of the United States (2001) Address to the Nation on Stem Cell Research from Crawford Texas, 37 Weekly Compl. Pres. Doc. 1149 (9 August 2001).

The President of the United States (2007) Executive Order No. 13435, 72 Fed. Reg. 34,591 (20 June 2007).

The President of the United States (2009) Executive Order 13505 of March 2009. Available at http://edocket.access.gpo.gov/2009/pdf/E9-5441.pdf

UNESCO (1997) *Universal Declaration of Human Genome and Human Rights.* Available at http://portal.unesco.org/en/ev.php-URL_ID= 13177&URL_DO= DO_ TOPIC&URL_SECTION= 201.html

UNESCO (2005a) *Explanatory Memorandum on the Elaboration of the Preliminary Draft Declaration on Universal Norms on Bioethics.* Paris: UNESCO.

UNESCO (2005b) *Universal Declaration on Bioethics and Human Rights.* Available at http://www.unesco.org/new/en/social-and-human-sciences/themes/bioethics/ bioethics-and-human-rights/

US: Guidelines for the Conduct of Research Involving Human Subjects (2004) Available at http://ohsr.od.nih.gov/guidelines/GrayBooklet82404.pdf

US: Research involving human subjects, Common Rules, Code of Federal Regulations, Title 45, Part 46. Available at http://www.hhs.gov/ohrp/refernces/ comrulp2.pdf

8
Governing Innovation Paths in Regenerative Medicine: The European and Global Struggle for Political Advantage

Brian Salter

Introduction

Over the last decade, 'innovation' has acquired an iconic status in the pantheon of state policies as governments compete for access to the knowledge economies of the future through a search for the appropriate alchemy of innovation governance. Propelled by the imperatives of globalisation, the expectations of their populations and the geopolitics of inter-state competition for future economic territories, ambitious governments have uniformly come to regard innovation policy as the key to unlocking the potential offered by the advancement of science. With the advent of the emerging economies of the developing world, we see an added political impetus as countries such as China, India, and Brazil have aggressively moved to establish their own innovation platforms. In their turn, the developed countries of North America, Europe, and Japan are well aware that they must respond to the challenge posed by the emerging economies to their traditional leadership of scientific innovation.

Nowhere is this dynamic more apparent than in the life sciences and regenerative medicine where the promise of future health, wealth, and happiness forms a staple part of the political narrative. This chapter examines how the consequent global competition for political advantage in regenerative medicine innovation has intensified the production of new forms of governance designed to enable states and regions such as the European Union (EU) to compete more effectively. Governance has become a knowledge terrain in its own right, fuelled by the political demand that it should constantly reformulate itself to accommodate the requirements of scientific and technological innovation. In the global context, control of governance production and governance territory is increasingly an integral part of the political game for innovation advantage.

With the advent of the governance knowledge market, a range of players have emerged anxious to maximise their influence over the operation of this market and the power embedded in it. Not only nation states but multinational corporations, international non-governmental organisations, epistemic communities, national regulatory agencies, and other transnational networks seek to engage, directly and indirectly, with the emergence of new governance knowledge. In the case of the governance of innovation in regenerative medicine, these actors may be concerned with a wide range of policies designed to impact on the governance domains of science, society, and the market: public or private governance interventions in terms of support for the science, mediation between the science and its cultural context, the maintenance of consumer confidence through the regulation of some or all of the stages of knowledge production, and the stimulation of market interest through intellectual property regulation, venture capital support, and public–private partnerships.

In aggregate, and depending on how the policy components are combined by particular states, the combination of these governance domains constitute competing models of biomedical innovation that will benefit some and disadvantage others. Advanced by different states, or groupings of states, they represent competing perspectives on the political economy of innovation. Historically it is the US model that has prevailed in the global 'value chain' of the life sciences, supported by Europe and Japan. The task of this chapter is to determine whether this hegemony is likely to prevail in the face of the challenges from the emerging economies.

Governing the political economy of innovation

Innovation governance in the life sciences can be characterised as a contested political terrain because it is the site for the struggle for future markets and future wealth at the state, international, and global levels. The significance of governance is that it defines what, where, and how innovation takes place through the imposition of rule systems to guide behaviour. These rule systems are not neutral but, as politically constructed entities, are designed to promote and protect the interests of one group of political actors over another. They constitute the building blocks of the political economy of innovation. It is therefore important for any state or other agency wishing to play an active part in the future of the bioeconomy to have the capacity to engage in the production of governance knowledge that will enable them to promote, defend, and negotiate their own interpretation of innovation. Lacking such a knowledge base, they become vulnerable to dominance by the rules produced by others.

The production of governance knowledge takes place in parallel to the production of scientific knowledge: both are necessary if the progress of a concept from scientific idea to marketable product is to occur. In the case of the life sciences, the scientific knowledge production process from the basic science, through clinical experimentation and trials, to the therapeutic product is long, arduous, and uncertain. At all stages in that process, there exists a potential triangle of tensions between the primary components of the socio-economic context of such knowledge production: the science may prove to be inadequate, society unsympathetic, or the market uninterested. As a result there is political pressure for governance to be 'co-produced' with science in order to respond to and if possible resolve these tensions. As Jasanoff puts it, this means a focus on how 'knowledge making is incorporated into practices of state-making, or of governance more broadly, and in reverse, how practices of governance influence the making and use of knowledge' (Jasanoff, 2004, p. 3).

From the perspective of the state, governance choices about the *science* have much to do with the creation and husbanding of the resources necessary for the enterprise to have an explicit domestic platform. This may require investment, an adequate research funding market, organisation of the scientific effort, and an appropriate supply of scientific labour and research materials such as oocytes (see Chapter 5) and stem cell lines. Secondly, regardless of the type of political system, the response of *society* to biomedical science may require governance choices to be made about how that response is negotiated both domestically and internationally if public trust in the field is to be maintained. Even if, as in China, the public voice is muted, both elite and international opinion, nonetheless, act to request, if not require, policies that at least appear to regulate the science in the public interest – in terms of not only risk and safety but also the sensitivities of cultural values. Without such policies, future consumer demand may be fatally undermined. Finally, the risk of *market* failure during the long gestation from basic science to eventual therapy means that early government funding intervention may be

Table 8.1 Governance choices in the political economy of innovation

Science
Investment in, and organisation of, the science
The training, retention and, if necessary, acquisition of the scientific labour
  necessary for the required knowledge production to take place

Civil society
Maintenance of public trust in all stages of the knowledge production process
  (laboratory, animal testing, clinical experimentation, clinical trials,
  commercial production)
  Culture – The cultural acceptability to citizens of the aims, conduct, and
    materials of the basic science and, in the event of cultural conflict, the
    regulation required to ensure compatibility with the dominant social
    values
  Safety – The protection of citizens, maintenance of consumer confidence,
    and the integrity of the potential product

Market
Ownership of the new intellectual property: the balance to be struck between
  the needs of the knowledge market, the freedom of science to access research
  results, and the cultural status of the new knowledge
Stimulation of the market response through support for the venture capital
  function, public–private partnerships, and pharma engagement

necessary to motivate patenting, venture capital investment, and pharmaceutical engagement in an emerging industry. Table 8.1 summarises the governance choices available to states wishing to compete in the future of the bioeconomy (Salter and Faulkner, 2011).

Science, society, and the market can be construed as general policy domains where governance interventions may be required in order to maintain the impetus of knowledge production. However, in the globalised world of the life sciences the construction of global advantage by a state is also obliged to recognise that in the operation of these domains: (a) the national–international levels of governance play different roles which may be complementary or conflictual; (b) direct public intervention may be counterproductive and at the very least need to be matched by indirect policies aimed at stimulating private-sector involvement; and (c) private governance may, depending on the political problem being addressed, be a more appropriate mode to adopt. This is because the dynamic of knowledge production in the life sciences is underpinned by a set of interlinking national and transnational markets: the funding market of scientific research, the scientific labour market, the moral economy of ethics for the trading of regulatory values, the intellectual property market, and the financial venture capital market – all energised by the political competition for advantage.

What this means is that in pursuing their desired political economy of innovation, states not only are frequently working across jurisdictions in a multi-level governance system of one kind or another (Hooghe and Marks, 2003) but are also obliged to recognise that the complexities of scientific advance and the need for continuous technical and ethical rule making have established a realm of private governance with its own networks, authorities, and procedures (Knill and Lehmkuhl, 2002; Büthe, 2004). As a result, states face a highly complex engagement with hybrid modes of governance requiring constant flexibility and adaptation if they are to keep up with the leaders (McGuinness and O'Carroll, 2010).

States and the global dynamic in the political economy of innovation

In deciding how to engage with the multi-dimensional character of governance production in the political economy of innovation, states are simultaneously driven by a common belief in the future wealth of a particular field of the life sciences and an awareness that to be dilatory is to lose position in the global competition for advantage. This powerful combination of carrot and stick has produced a preoccupation with 'innovation' as a guiding leitmotif in economic policymaking. It is an edgy global contest. From those states of the developed world which have traditionally dominated access to new markets through the exclusive production of new knowledge comes a political rhetoric laced with the fear of losing the innovation race. In its first annual report in 2011, the EU's Innovation Union initiative begins:

> The shift in economic power from West to East is accelerating. Both the Innovation Union Scoreboard and the Innovation Union Competitiveness report highlight the fact that Europe's research and innovation performance has declined over recent years, causing a broadening of the already sizeable innovation gap vis-à-vis the US and Japan. Furthermore, China, India and Brazil have started to rapidly catch up with the EU by improving their performance 7%, 3% and 1% faster than the EU year on year over the last five years.
>
> (European Commission, 2011)

In the United States, the *American strategy for innovation* (2011) observes that 'across a range of innovation metrics...our nation has fallen in global innovation-ranked competitiveness.' It continues, given that 'America's future economic growth and international competitiveness depend on our capacity to innovate.... To win the future, we must out-innovate, out-educate, and out-build the rest of the world' (Executive of the President, 2011). The urgency of the situation is reinforced by the data. Commenting on the publication of the Science and Engineering Indicators 2012 and its data on the rapidly increasing Asian investments in knowledge-intensive economies, the National Science Foundation (NSF) Director Subra Suresh observed, 'This information clearly shows we must re-examine long-held assumptions about the global dominance of the American science and technology enterprise' (National Science Foundation, 2012).

The examination of the challenge to the Western hegemony in innovation is accompanied by an awareness of the difficult issues that flow from the transnational character of innovation. In *Innovation nation*, UK policymakers recognise both that the 'emerging economies in particular are likely to challenge us for our position in the future' and that such economies may form part of the solution given the international trend towards 'open innovation' where new knowledge and ideas are shared, commercialised, capitalised, and traded (Department of Innovation, Universities and Skills, 2008, p. 42). The implication is that 'many of the major policy challenges identified in the coming decades arise from global problems [of innovation] that will require global collaborations to deliver solutions' (Department of Innovation, Universities and Skills, 2008, p. 50). Similarly, the examination of 'innovation eco-systems' in the Sainsbury Review of Science and Innovation Policies, optimistically entitled *The race to the top*, recognised the dispersed and complex nature of innovation value chains and highlighted that the 'capability to integrate stages globally may be a major opportunity for the UK to draw on its traditional strengths in innovation and its international outlook' (Lord Sainsbury of Turville, 2007, p. 42).

Whilst the concern of states of the developed world is how best to maintain their leadership position in innovation, that of the emerging economies is with how best to challenge it. Having successfully played 'catch up' through the creation of industries that enable them to compete in established global markets with a known demand, states such as China, India, South Korea, and Taiwan have embraced the political narrative of innovation as the necessary condition for maintaining their upward economic trajectory to the markets of the future where the demand is unknown (Kim, 1997; Weiss, 2003). This strategic shift, and the accompanying symbolism of the innovation rhetoric, is exemplified in President Hu Jintao's landmark speech at the opening of China's Fourth National Conference on Science and Technology in January 2006 in which he strongly emphasised China's need to 'adhere to a new path of innovation with Chinese characteristics [zi-zhu-chuang-xin] and strive to build an innovation-oriented country' (Gov.cn, 2006). His message was clear. In a rapidly changing global market, China could no longer rely on the economic advantages afforded by a cheap labour force which exploited the inventions of others. Instead, if it is to retain its international competitive advantage in the context of the economies of the developed world, China must rapidly develop an indigenous science and technology platform with the capacity to establish its own innovative directions, exploit its own intellectual capital, and establish its own new industries. President Hu's message was subsequently incorporated into the 11th Five Year Plan 2006–2010 and the range of policies flowing from it, including biomedicine and the development of new health technologies. Similarly, in 2010 the Indian president announced the launch of a 'Decade of Innovation' to be taken forward by a new National Innovation Council guided by a strategy based on the principle that 'the future prosperity of India in the knowledge economy will increasingly depend on its ability to generated new ideas, processes and solutions, and through the process of innovation convert knowledge into social good and economic wealth' (Office of Adviser to the Prime Minister, 2011, p. 3). Also, like China, India sees itself as creating a distinctive and, in this case, communitarian approach to innovation (the 'Indian Model of Innovation') geared to the particular social and economic needs of developing countries: 'India needs more "frugal, distributed, affordable" innovation that produces more "frugal cost" products and services that are affordable by people at low levels of income without compromising the safety, efficiency, and utility of such products' (National Innovation Council, 2011).

The commitment of states to innovation strategies geared to future, and unknown, markets is underpinned by the promise of science that particular knowledge domains can be developed to the point where commercialisation can take place. Without this promise, and without the trust in the promise that the authority of science can command, the global politics of innovation would not work. Politicians and policy-makers have to persuade both themselves and their constituencies that the transition from the current science to the future product is possible. For the innovation process to be sustainable, they must maintain the imaginary future world as viable and achievable.

This exercise in political imagination begins with the delineation of the future value that a particular area of science can deliver. In the case of regenerative medicine, the future value lies in its claimed capacity to produce therapies for a broad range of diseases and conditions including diabetes, heart disease, renal failure, osteoporosis, and spinal cord injuries for which there is at present only partial treatment or none at all. Although almost none of the promise of regenerative medicine has been realised, in a political sense this does not matter if the capture of the imaginations of publics, patients, or politicians can be maintained over time. Embedded in these imaginations are hopes and expectations of what the future might bring and, if the faith is sufficiently strong, a commitment to support the allocation of the resources required to enable that imagined future to become reality (Brown et al., 2000; Brown, 2003; Brown and Michael, 2003). In power terms, it is this political sustainability that is important, not whether the belief subsequently turns out to be true. Recent examples such as transgenics, reproductive science, and bioinformatics reveal the competitive nature of this enterprise and the importance of not allowing a rival area of science to capture the imaginative high ground. For if public support is gained then the authenticity of the future market becomes more tangible; if political support is gained then the winning of scarce scientific resources becomes more likely. At the same time, scientific advocates must beware of over-hyping their future products since this may overstretch the imaginary envelope and cause both its collapse and that of its associated anticipated future values. To be politically effective, advocates must be seen to act responsibly, rationally and with due discretion. They must also be entrepreneurial since given the uncertainty of the regenerative medicine field and the long gestation from the basic science to the anticipated product, its political imagination cannot afford to be static but must constantly evolve if its promise is to be maintained and its doubters diminished. Hence we find that as a scientific concept 'regenerative medicine' flexes in the face of new demands, new epistemic partners, and new possibilities, facilitated by the promissory politics of entrepreneurial scientists with a stake in the survival and development of the field (Morrison, 2012).

Ably supporting the scientific contribution to the political imaginary of regenerative medicine is a small industry of economic forecasters using a variety of techniques to predict the future size of the market. Although the predictions vary wildly, there, nonetheless, is a common confidence that regenerative medicine is a substantial future industry. In 2010, for example, a report by Global Industry Analysts Inc.

forecast a future global regenerative medicine market of US$1.4 billion by 2015 (Global Industry Analysts Inc., 2010). Such reports are not discouraged by the predictive failure of their predecessors. Thus a 2006 US government report 'conservatively estimates' the worldwide market for regenerative medicine to be US$500 billion by 2010 (US Department of Health and Human Services, 2006), while a British Standards Institute report announced that the European market was expected to reach US$15 billion by the same year (British Standards Institution, 2006). Forecasts by consultancy firms on the stem cell market include a US market of US$3.6 billion and a world market of US$8 billion by the same year. Others were less optimistic, predicting a world market of US$100 million by 2010 rising to US$2 billion by 2015 (Biophoenix, 2006). Although the market forecasting of regenerative medicine is a highly variable exercise dependent on what diseases, technologies, and types of firm are included, its contribution to the political imaginary is the consistent message that the economic future of regenerative medicine is real and achievable even though the precise nature of this reality may be elusive. This contribution is assisted by the character of the forecasting which is usually detached from any consideration of the impact on patient demand for the new health technologies of factors that would question the viability of the future market such as the mode of health care funding, cost constraints, ethical considerations, regulatory frameworks, or policies aiming at the protection of domestic pharmaceutical industry.

The global capture of states by the political imaginary of regenerative medicine is evidenced by its inclusion in the priority innovation agendas of all states with the ambition of competing in the future economy of the life sciences. Unsurprisingly, its inclusion is often accompanied by justifications that reflect the hegemonic understandings of innovation policy in general. Thus the 2006 US Department of Health and Human Services report on regenerative medicine starkly warned the nation that 'the US pre-eminence in the field of regenerative medicine is in jeopardy' unless appropriate investment and support is provided (Department of Health and Human Services, 2006, p. 13). Similarly, the United Kingdom's 2005 report on a ten-year strategy for the development of stem cell research, therapy, and technology (the 'Pattison Report') made clear its ambition 'for the UK to consolidate its current position of strength in stem cell research and mature . . . into one of the global leaders in stem cell therapy and technology' (UK Stem Cell Initiative, 2005, p. 5). The report was at pains to document in detail the investments being made by other competitor countries in stem cell science. One of these, Germany, produced a study that whilst noting Germany's leading international position in regenerative medicine argued for urgent policy changes to facilitate the translation from the science to the therapeutic product (German Federal Ministry of Education and Research, 2007).

Anxious not to be excluded from future markets, states wishing to challenge the hegemony of the West in the development of health technologies have adopted a similarly positive vision of regenerative medicine science with China, India, South Korea, and Singapore all making substantial policy commitments to the field (Padma, 2005; Kim, 2006; Holden and Demeritt, 2008; Lander et al., 2008; MCMahon et al., 2010). Thus, in February 2011, the Chinese Academy of Sciences (CAS) unveiled its *Innovation 2020* with a set of seven new research priorities, including regenerative medicine, designed to facilitate the contribution of scientific innovation to economic success. As part of this initiative the CAS plans to establish 'a world-class research platform and base for the study of stem cell and regenerative medicine encompassing four research centres in Beijing, Shanghai, Guangzhou and Kunming and leveraging the resources of 17 other institutions around the country' (Chinese Academy of Sciences, 2011).

Hegemonic challenge and state adaptation

In very general terms, the nature of the global competition for advantage in innovation is reflected in the science governance domain where R and D investment can impact on the innovation infrastructure. In 2009, the current state of play shows that global R and D performance was concentrated in three geographic regions: North America (United States, Canada, Mexico – US$433 billion, 34% of total), Europe (US$319 billion, 25%) and Asia (China, India, Indonesia, Singapore, Malaysia, Taiwan, Thailand, Singapore, Japan, South Korea – US$402 billion, 32%) (Figure 8.1). Three countries account for more than half of global R and D: the United States (US$402 billion, 31% of total), China (US$154 billion, 12%), and Japan (US$138 billion, 11%). Rather more important are the underlying trends and the immanent shifts in the hegemonic order. In the last ten years Asia's share of global R and D has increased from 24% in 1999 to its current 31% underpinned by a dramatic rise in China's contribution which has averaged an annual increase of about 20% (28% in 2008/2009). In contrast to this, whilst the United States retains its dominant global status, its pace of growth in R and D performance over the same period has averaged 5% with a consequent decline in its relative position from 38% in 1999 to 31% in

Dollars (billions)

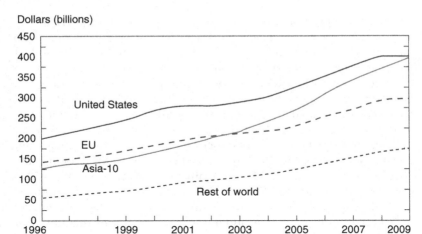

Figure 8.1   R&D expenditures for the United States, EU, and ten Asian countries

2009. Europe exhibits a similar pattern with an average growth of 5.8% and a decline from 27% to 25% over the same period. For the future, it is interesting to note that although China's R and D expenditure expressed as a proportion of GDP has tripled since 1996, it is still only 1.7% compared to the US figure of 2.88%, and therefore likely to increase further (National Science Board, 2012, pp. 0.4–0.5).

Within the science governance domain, a second key indicator of changes in innovation capacity is researcher workforce and outputs – with a similar global pattern emerging. Whilst US and EU growth rates averaged at or below 3% and 4% between 1995 and 2009, that of the Asian region outside Japan averaged 8%–9% for Taiwan, Singapore, and South Korea with China averaging 12% in the 2002–2009 period (National Science Board, 2012, p. 0.9). Although researchers in the EU and the United States have long dominated world article production their combined share of published articles decreased from 69% in 1995 to 58% in 2009. Meanwhile, Asia's world article shared expanded from 14% to 24% in the same period with China's annual growth averaging 16%. As a consequence, by 2007 China had moved into second place behind the United States. There are clear differences emerging in the priority fields supported by states. For example, whilst a large proportion of US articles focused on the biomedical and other life sciences, outputs from Asia tended to concentrate on the physical sciences and engineering (National Science Board, 2012, p. 0.10).

In the face of this challenge, both individual states and the supranational government of the EU have sought to adapt their modes of governance to the uncertainties inherent in the innovation process. Whilst in some cases this has meant the introduction of new forms of governance, in the developed economies it has also led to the refinement and consolidation of what are regarded as the tried and tested approaches to innovation which have served the hegemony well in the past. Although there are differences between Western states in terms of the specifics of implementation, they have largely moved away from the national sponsorship of particular firms and technologies and towards policies that foster an infrastructure supportive of the conditions necessary for innovation. Such an infrastructure, it is held, is better able to respond dynamically to the inherent uncertainties of innovation. Customary governance components of this approach are the maintenance of existing research funding; initiation by regional governments of programmes to foster cluster developments in sectors such as biotechnology (Asheim and Gertler, 2004); the facilitation of commercialisation through close academic–industry collaborations and high-profile, publicly funded R and D centres acting as magnets for venture capital investments (Cooke, 2003, 2004); facilitative rather than restrictive regulation (Hansen, 2001); and patenting arrangements that favour the operation of the market.

In the United States, in terms of the governance of the science, any challenge to its hegemony in regenerative medicine has to be placed in the context of its steadily expanding commitment to research in the life sciences. Over the last 20 years, the balance of R and D expenditure has shifted steadily away from the physical sciences and in favour of the life sciences so that in 2009, of the total federal funding for basic and applied research of US$63.7 billion, US$33.3 billion (52%) supported research in the life sciences (National Science Board, 2012, p. 4.33). During this period, the life sciences were the only field to experience an expansion (6%) of their share of the total academic R and D (National Science Board, 2012, p. 5.4). At the same time, life sciences accounted for much of the growth in the academic science and engineering doctoral workforce so that in 2008 life scientists represented more than a third of such researchers (National Science Board: 5.50). In the field of regenerative medicine the commitment is particularly evident: in the 2008–2011 four-year period, the National Institutes of Health invested US$3.2 billion in regenerative medicine research and US$4.2 billion in stem cell research (classified separately) (National Institutes of Health, 2012). US states add to this total through their own funding schemes so that in the two years leading up to 2008 a further US$4.5 billion was committed to the field as a result of inter-US state competition for advantage in regenerative medicine (National Conference of State Legislatures, 2012).

No other country remotely approaches these levels of funding in the life sciences and regenerative medicine. Where a specific commitment to regenerative medicine is made, be this in Europe or the emerging economies of Asia, it tends to be less than US$100 million over, typically, a two- to five-year period (Salter, 2009a). In the United States, the formative power of this initial scientific investment is reinforced by the market-oriented governance of innovation. The American approach is typical of the 'triple helix model' where there is a continuing iteration between the knowledge-producing sector (university science), the market, and government. Continuing and cyclical interactions between the relevant agents are seen as more appropriate to the innovation-based knowledge economy rather than one based on an assumption of linear development (Etzkowitz and Leydesdorff, 1997, 1998, 2000). Here the role of governance at both the federal and US state level is to facilitate

> the incubator environments that academic research institutions can provide, the availability of venture capital and professional business advice, exit opportunities for early investors, the possibilities for the diffusion of knowledge and, last but not least, the competence of political actors to distribute research money where innovations are most probable.
>
> (Giesecke, 2000, p. 217)

It is in this kind of 'new economy innovation system', as Cooke (2001) describes it, that venture capital is able to act to facilitate university–industry interactions in areas of uncertainty. Venture capitalists (VCs) contribute not only through their investment in high-risk, early-stage scientific ventures such as regenerative medicine but also through their management involvement in the successful commercialisation of such ventures. Here the governance role of the state is to create a supportive environment through such measures as the 1980 Bayh–Dole Act that allowed universities to benefit commercially through the exploitation of their inventions. When taken together with the Small Business Act of 1958 it shows how the state enabled the development of an innovation model where venture capital companies could act as an equity and investment market capable of supporting biotechnology development. The success of this model in sustaining US dominance in the life sciences is reflected in the pattern of global venture capital investment. In the life sciences in 2010, US venture capital of US$6168 million constituted 71% of the global total of US$8649 million, rising to 76% (US$6998 million) of the total US$9230 million in 2011 (Burrill Report, 2012).

The US model is characterised by a pluralist and heterogeneous approach to biomedical innovation – witness the absence of a US national agency for science and technology policy. Commentators have contrasted this with the approach of European countries such as Germany where discrete policy silos help to reinforce a lack of institutional interaction between universities and industry with the consequent implication that the emergence of a triple helix model is severely constrained (Giesecke, 2000). The EU's awareness of and response to the limitations of its members' governance of innovation is exemplified in the 2000 Lisbon Strategy for the EU 'to become the most competitive and dynamic knowledge-based economy in the world'. The Strategy envisaged the use of a range of governance instruments that included resource distribution through a new European Area for Research and Innovation and the redirection of the European Investment Bank and the European Investment Fund towards providing support for high-tech firms; legal measures for harmonisation such as a Community Patent; and a new form of network governance through the new Open Method of Coordination (OMC) (De La Porte, 2002; Regent, 2003). It was the latter that attracted much attention since it seemed to promise a shift away from the rigidity of traditional modes of support for innovation towards a more flexible and horizontal form of governance. The OMC had several phases: the establishment of goals and guidelines by the European Council, their elaboration by the Council of Ministers through specific targets and benchmarks, indicators and timetables, the development of policies by member states, and their evaluation by the Commission and Council. Its networks were to comprise not only EU institutions and member states but also sub-national authorities and civil society organisations and industry.

In practice, the utility of the OMC as a governance mechanism for supporting and promoting EU innovation in regenerative medicine has proved limited, with the use and adaptation of established mechanisms a prominent characteristic of the regenerative medicine field. In the governance domain of science, the EU's Framework Programmes have provided continuing support. In FP6, 111 projects totalling 532 million euros involved stem cells and, in FP7, by May 2010 a total of 187 million euros had been committed to regenerative medicine projects (European Commission, 2009; Kessler, 2010). Meanwhile, the EU's Structural Funds have been drawn on by researchers as a source for funding the physical infrastructure of regenerative medicine. For example, regenerative medicine researchers at the Czech Republic's Central European Institute of Technology, Germany's Rostock Stem Cell Therapy Centre, and the Scottish Centre for Regenerative Medicine have all received infrastructure funding (Salter and Hogarth, 2011, p. 8).

A similar picture of governance continuity is apparent in the domains of society and market governance where regulation is traditionally employed to deal both with the maintenance of public trust (health and safety, cultural issues) and the harmonisation of standards across disparate market locations (Waldby and Salter, 2008). EU regenerative medicine products are now regulated under Regulation 1394/2007 on Advanced Therapy Medicinal Products. In large part the impetus for an EU-wide approach to the regulation of regenerative medicine came from industry concerns that wide variations in regulation across member states had created a 'heterogeneous and segmented market in Europe' (Hughes-Wilson and Mackay, 2007). Prior to the 2007 regulation, regenerative medicine products were regulated by different member states as devices, drugs, biologics or a combination of them, or dealt with specifically as cell therapy products. The 2007 regulation brought them under the framework of pharmaceutical regulation with authority for product approval vested in the European Medicines Agency (EMA). The success in centralising product approval under EMA is consonant with the traditional view of the EU as a 'regulatory' state with particular strengths in the production of regulatory policy (Majone, 1996).

Somewhat ironically, where the OMC governance style has found traction in the field of regenerative medicine it has been through the already established infrastructure of innovation support. Thus the Framework Programmes have facilitated transnational scientific networking through the funding of such projects as the human embryonic stem cell registry (hESCreg) that contribute to collaborative standard setting in regenerative medicine research, engaging with such entities as the International Consortium of Stem Cell Networks. HESCreg acts as a vehicle for transnational governance by promoting technical and ethical standards supported by a validation process to ensure that registered stem cell lines comply with its requirements. It is likely, therefore, that in regenerative medicine, as elsewhere, the new modes of innovation governance have not become central despite what Kassim and Le Gales term 'the revolt against traditional methods and the backlash against regulation' – 'governance has not systematically replaced government' (Kassim and Le Gales, 2010, p. 14). Rather, the promotion of OMC processes through FP projects illustrates the potential of what McGuinness and O'Carroll term 'hybrid modes of governance... a mix of Community action and non-legislative governance', which they suggest might offer 'the most potential for the future of OMC' (McGuinness and O'Carroll, 2010, p. 312).

Whereas the EU faces the conundrum of the relationship between old and new forms of innovation governance, in the absence of an historic innovation capacity the emerging economies have had to confront the issue of where to place their efforts as they seek to build their innovation governance infrastructures from scratch. In so doing, they have had to adapt their developmental state approach which traditionally was founded on the promotion of rapid economic development through the targeting of particular industries with known global markets (Onis, 1991; Hawes and Liu, 1993). The investment–return relationship was very tangible. The role of the state was to protect their selected industries through policies such as import and credit controls and promote them through direct state investment and the guidance of private capital (Wade, 1990). Backed by a professional bureaucracy, the state aimed to define the path of industrialisation through 'government of the market'. However, this mode of governance does not work in biomedical innovation where the future market is largely unknown and the route to commercialisation characterised by the uncertainty of the innovation process itself. So although general statements can be made about future regenerative medicine products, precisely what they will look like, how they will work, what range of disciplines are required for their realisation, and how long they will take to achieve are questions to which there are no answers. Governance therefore becomes a vehicle for the management of uncertainty through the creation of general capacities and processes rather than the identification of specific goals. Here the investment–return relationship is very intangible.

As we have seen, the governance of science in the emerging economies is characterised by very rapid increases in R and D expenditure, the scientific workforce, and publications. Although the overall balance of this commitment favours the physical sciences and engineering, some states have chosen to invest heavily in the life sciences and the perceived promise of health technologies. In Taiwan and Singapore, for example, the first decade of this century has seen allocations on the life sciences of approximately one-third of government total R and D expenditure (Wong, 2011, p. 5). However, the necessary governance of uncertainty associated with the development of new health technologies has left the emerging economies in an exposed position given the failure of biotechnology in general and regenerative medicine in particular to produce benefits in any sense commensurate with the investment costs incurred. The first wave of regenerative medicine has had limited clinical adoption and a high level of commercial failure compounded by a lack of clear business models, fragmentary clinical organisation, and the challenge regenerative medicine poses to existing procedures for product licensing given its uncertain regulatory classification (device; cell therapy; pharmaceutical) (Martin et al., 2009). As Wong comments of biotechnology, 'poor performance of the sector has provided so little clarity about the sector that decision makers are not able to get any feel for the probability of success, and they have even less sense of how to increase that probability' (Wong, 201: 32).

In this situation, whereas states of the developed world can fall back on their established infrastructure of innovation, adapt it here, and enhance it there whilst tolerating uncertainty, states of the developing world are not in this position. Accustomed to making choices, picking winners, reaping the rewards, and moving on to the next project, such states have taken the first steps through commitments in the governance of the life sciences but remain unclear as to what supporting measures should be taken in the other governance domains of society and the market. In the domain of society governance, there is an awareness that, in response to international as much as national public opinion, regulation is needed to address both health and safety and cultural concerns surrounding new health technologies such as regenerative medicine. A common strategy is to make a highly visible national commitment to the regulation of, for example, stem cell science, but then be unable to implement the policy effectively in the absence of established infrastructures for the support of this type of governance – as in the case in China and India (Salter and Qiu, 2009). Rather more success has been achieved in the market governance domain where World Trade Organisation (WTO) membership has imposed certain governance conditions, notably the introduction of robust patenting regimes, though again there are issues associated with the thoroughness with which such a relatively new system of governance can be successfully enforced in market cultures which have thrived on their absence (Salter, 2011). State support for the use of venture capital in innovation is also evident but as in China, for example, it is not accompanied by the skills necessary for assessing the risks inherent in the development of new health technologies such as regenerative medicine (Salter, 2009b). Without such skills, investment in high-risk health technology ventures will either fail because of an inability accurately to estimate the market potential or not occur at all. The result is very low VC investment in such fields accompanied by a very limited capacity to penetrate the health care markets of the West, particularly that of the United States. Patents are a prerequisite and important measure of such penetration and the most recent US Patent and Trademark Office data show that between 1992 and 2010 China's share of awards to non-US inventors rose from below 0.5% to 3%, scarcely a global threat (National Science Board, 2012, p. 0.13).

Conclusion

The particular nature of innovation in the life sciences raises strategic issues for states wishing to compete in the future global bioeconomy. Innovation in this field is long, arduous, and inherently uncertain in terms of both process and outcomes. This, combined with the scale of the resources required, means that it inevitably incurs large opportunity costs. Yet at the same time the vision of future benefits and wealth that the life sciences conjure is pervasive, powerful and, for many ambitious states, irresistible. In particular, the political imaginary of regenerative medicine is manifest not only in the statements of leading politicians but also in the policy priorities of their governments. To varying degrees, states have committed themselves to modes of governance designed to enhance their capacity for health technology innovation through infrastructure development in the domains of science, society, and the market.

The intention is that such governance interventions will improve a state's ability to position itself in the global competition for future advantage. However, any assessment of the value of these interventions must begin with the recognition that the competition does not take place on a level playing field. The overwhelming preponderance of the United States in the resources committed to the governance of the life sciences and regenerative medicine, and the trend for these resources to constitute an increasing proportion of its national R and D budget, means that when assessed in relative terms the challenge from the emerging economies is inevitably weak. Even though the absolute rate of R and D growth of these states is impressive, it is much less so in the specific case of the life sciences. They may have 'caught up' with the developed world in the known industrial markets but the completely different economic and political costs of speculative innovation in the unknown markets of the future bioeconomy mean that catch-up in this sector for the most part remains a distant dream.

Even developed world states accustomed to dealing with the problematic of life sciences innovation have struggled to find forms of governance that can cope with the long-term uncertainties of the field. In Europe, the attempts to introduce innovation-friendly modes of network governance in the wake of the Lisbon Strategy have largely failed to impact in areas such as regenerative medicine. Instead, there have been various forms of adaptation of the existing governance modes of resource distribution and regulation that have gone some way to stimulate innovation potential. To an extent this can be regarded as a form of governance path dependency where the bureaucratic dynamic of existing modes of governance militate against the acceptance of new modes that are founded on different operational principles.

There is a final question regarding the political limits of a state's ability to tolerate continued uncertainty in the regenerative medicine field of innovation. Given the absence of clear benefits, the complexities of continuing governance invention, and the opportunity costs incurred, at what point will governments be obliged to align the political imaginary of regenerative medicine with the reality of their limited resources and choose to exit the global competition? As Wong has noted, states in the developing world have sought to mitigate the political risks of biotechnology innovation by retreating from direct forms of governance, by becoming less overtly 'developmental' and more facilitative in their interventions and, as a result, similar in their governance arrangements to the states of the developed world (Wong, 2011, pp. 179–181). This has bought them time but is not sustainable in the long term. The difficulty of course is the political costs of exit from a lengthy, expensive, and highly visible strategy where innovation has been explicitly linked to national prestige. The honest admission of failure has few political attractions.

References

Asheim, B. T. and M. S. Gertler (2004) The geography of innovation: regional innovation systems, in J. Fagerberg, D. C. Mowery and R. R. Nelson (eds) *The Oxford Handbook of Innovation*. Oxford: Oxford University Press, pp. 213–237.

Biophoenix (2006) *Opportunities in Stem Cell Research and Commercialisation*. London: Business Insights, p. 80.

British Standards Institution (2006) *Guidance on Codes of Practice, Standardised Methods and Regulations for Cell-based Therapeutics from Basic Research to Clinical Application*. PAS 83. London: British Standards Institution. Available at: http://www.bsi-global.com/en/Standards-and-Publications/Industry-Sectors/Healthcare/Regenerative-Medicine-/?recid= 819, accessed 6 February 2012.

Brown, N. (2003) Hope against hype: accountability in biopasts, presents and futures, *Science Studies*, 16(2): 3–21.

Brown, N. and M. Michael (2003) A sociology of expectations: retrospecting prospects and prospecting retrospects, *Technology Analysis and Strategic Management*, 15(1): 3–18.

Brown, N., B. Rappert and A. Webster (eds) (2000) *Contested Futures: A Sociology of Prospective Technoscience*. Aldershot: Ashgate.

Burrill Report (2012) Biotech sector scores wins in November. Burrill Report. Available at: http://www.burrillreport.com/article-the_january_2012_issue_of_the_burrill_report.html, accessed 22 February 2012.

Büthe, T. (2004) Governance through private authority? Non-state actors in world politics, *Journal of International Affairs*, 58: 281–290.

Chinese Academy of Sciences (2011) Clean energy and stem cell research among China's science goals for 2020. Available at: http://english.cas.cn/Ne/CASE/201102/t20110209_64985.shtml, accessed 10 February 2012.

Cooke, P. (2001) New economy innovation systems: biotechnology in Europe and the USA, *Industry and Innovation*, 8(3): 267–289.

Cooke, P. (2003) The evolution of biotechnology in three continents: Schumpeterian or Penrosian? Editorial, *European Planning Studies*, 11(7): 757–763.

Cooke, P. (2004) Regional knowledge capabilities, embeddedness of firms and industry organisation: bioscience megacentres and economic geography, *European Planning Studies*, 11(7): 625–641.

De La Porte, C. (2002) Is the open method of coordination appropriate for organising activities at European level in sensitive policy areas?, *European Law Journal*, 8(1): 38–58.

Department of Universities, Innovation and Skills (2008) *Innovation Nation*. Cm 7345. London: The Stationery Office.

Etzkowitz, H. and L. Leydesdorff (1997) *Universities and the Global Knowledge Economy: A Triple Helix of University–Industry–Government Relations*. London: Pinter.

Etzkowitz, H. and L. Leydesdorff (1998) The endless transition: a 'triple helix' of university–industry–government relations, *Minerva*, 36: 203–208.

Etzkowitz, H. and L. Leydesdorff (2000) The dynamics of innovation: from national systems and 'mode 2' to a triple helix of university–industry–government relations, *Research Policy*, 29(2): 109–123.

European Commission (2009) *European Research Projects Involving Stem Cells in the 6th Framework Programme*. Available at: http://ec.europa.eu/research/fp6/p1/stemcells/pdf/stemcell_eu_research_fp6_en.pdf#view= fit&pagemode= bookmarks, accessed 22 February 2012.

European Commission (2011) *State of the Innovation Union 2011*. Report from the Commission to the European Parliament, the Council, the European Economic and Social Committee, and the Committee of the Regions, COM (2011) 849 final.

Executive of the President (2011) *American Strategy for Innovation*. Washington, DC: National Economic Council.

German Federal Ministry of Education and Research (BMBF) (2007) *Regenerative Technologies for Medicine and Biology – Contributions for a Strategic Funding Concept*. Berlin: Capgemini Deutschland.

Giesecke, S. (2000) The contrasting roles of government in the development of biotechnology industry in the US and Germany, *Research Policy*, 29(2): 205–223.

Global Industry Analysts Inc. (2010) Regenerative medicine: a global strategic business report. Summary available at: http://www.prweb.com/releases/regenerative_medicine/stem_cell_research/prweb4657624.htm, accessed 7 February 2012.

Gov.cn (2006) Chinese Government's official website. Available at: http://english.gov.cn/2006-01/09/content_151631.htm, accessed 6 February 2012.

Hansen, A. (2001) Biotechnology regulation: limiting or contributing to biotech development?, *New Genetics and Society*, 20(3): 255–271.

Hawes, G. and H. Liu (1993) Explaining the dynamics of the Southeast Asian political economy: state, society and the search for economic growth, *World Politics*, 45(4): 629–660.

Holden, K. and D. Demeritt (2008) Democratising science? The politics of promoting biomedicine in Singapore's developmental state, *Environment and Planning D: Society and Space*, 26(1): 68–86.

Hooghe, L. and G. Marks (2003) Unravelling the central state, but how? Types of multi-level governance, *American Political Science Review*, 97: 233–243.

Hughes-Wilson, W. and D. Mackay (2007) European approval system for advanced therapies: good news for patients and innovators alike, *Regenerative Medicine*, 2(1): 5–6.

Jasanoff, S. (2004) *States of Knowledge: Co-production of Science and Social Order*. London: Routledge.

Kassim, H. and P. le Gales (2010) Exploring governance in a multi-level polity: a policy instruments approach, *West European Politics*, 33(1): 1–21.

Kessler, C. (2010) EU support for stem cell research. Presented at: *EMA Workshop on Stem Cell-Based Therapies*. London, UK, 10 May 2010. Available at: www.emea.europa.eu/docs/en_GB/document_library/Presentation/2010/05/WC500090649.pdf, accessed 10 February 2012.

Kim, L. (1997) *Imitation to Innovation: The Dynamics of Korea's Technological Learning*. Boston: Harvard Business School Press.

Kim, T. G. (2006) $450 million budget set for stem cell research, *The Korean Times*, May 29.

Knill, C. and D. Lehmkuhl (2002) Private actors and the state: internationalisation and changing patterns of governance, *Governance*, 15(1): 41–63.

Lander, B. H. Thorsteinsdóttir, P. A. Singer and A. S. Daar et al. (2008) Harnessing stem cells for health needs in India, *Cell Stem Cell*, 3(1): 11–15.

Lord Sainsbury of Turville (2007) *Race to the Top: A Review of Government's Science and Innovation Policies*. London: HM Treasury.

Majone, G. (1996) *Regulating Europe*. London: Routledge.

Martin, P., R. Hawkesley and A. Turner (2009) *The Commercial Development of Cell Therapy: Lessons for the Future*. University of Nottingham: Institute for Science and Society.

McGuinness, N. and C. O'Carroll (2010) Benchmarking Europe's lab benches: how successful has the OMC been in research policy?, *JCMS*, 48(2): 293–318.

McMahon, D. S. et al. (2010) Cultivating regenerative medicine innovation in China, *Regenerative Medicine*, 5(1): 35–44.

Morrison, M. (2012) Promissory futures and possible pasts: the dynamics of contemporary expectations in contemporary medicine, *Biosocieties*, 7: 1–20.

National Conference of State Legislatures (2012) *Stem Cell Research*. Available at: http://www.ncsl.org/programs/health/genetics/embfet.htm, accessed 19 February 2012.

National Innovation Council (2011) *Report to the People*. National Innovation Council. Available at: http://www.innovationcouncil.gov.in/index. php?option=com_content&view=article&id=96:-report-to-the-people-2011& catid=8:report&Itemid=10, accessed 15 February 2012.

National Institutes of Health (2012) *Estimates of Funding for Various Research, Condition and Disease Categories*. Available at: http://report.nih.gov/rcdc/ categories/, accessed 19 February 2012.

National Science Board (2012) *Science and Engineering Indicators 2012*. Washington, DC: National Science Board.

National Science Foundation (2012) *New Report Outlines Trends in US Global Competitiveness in Science and Technology*. Press Release 12-011. Washington: National Science Foundation.

Office of Adviser to the Prime Minister (2011) *Towards A More Inclusive and Innovative India*. Indian Innovation Council. Available at: http://www. innovationcouncil.gov.in/images/stories/report/Innovation_Strategy.pdf, accessed 14 February 2012.

Onis, Z. (1991) The logic of the developmental state, *Comparative Politics*, 24(1): 109–126.

Padma, T. V. (2005) India plans stem cell initiative, *SciDevNet*, 13 January. Available at: http://www.scidev.net/News/index.cfm?fuseaction= readnews& itemid=1849&language=1, accessed 9 February 2012.

Regent, S. (2003) The open method of coordination: a new supranational form of governance?, *European Law Journal*, 9(2): 190–214.

Salter, B. (2009a) State strategies and the global knowledge economy: the geopolitics of regenerative medicine, *Geopolitics*, 14: 1–31.

Salter, B. (2009b) China, globalisation and health biotechnology innovation: venture capital and the adaptive state, *East Asian Science and Technology: An International Journal*, 3(4): 401–420.

Salter, B. (2011) Biomedical innovation and the geopolitics of patenting: China and the struggle for future territory, *East Asian Science and Technology: An International Journal*, 5: 1–18.

Salter, B. and S. Hogarth (2011) The dynamics of RM innovation in the EU: states, strategies and alliances, *REMEDiE Project*. Workpackage Four. Final Report.

Salter, B. and R. Qiu (2009) Bioethical governance and basic stem cell science: China and the global biomedicine economy, *Science and Public Policy*, 36(1): 47–59.

UK Stem Cell Initiative (2005) *Report and Recommendations*. London: Department of Health.

US Department of Health and Human Services (2006) *2020: A New Vision – A Future for Regenerative Medicine*. Washington: US Department of health and Human Services.

Wade, R. (1990) *Governing the Market: Economic Theory and the Role of Government in Asian Industrialisation*. Princeton: Princeton University Press.

Waldby, C. and B. Salter (2008) Global governance in human embryonic stem cell science: standardisation and bioethics in research and patenting, *Studies in Ethics Law and Technology*, 2(1): 1–23.

Weiss, L. (2003) Guiding globalisation in East Asia: new roles for old developmental states, in L. Weiss (ed) *States in the Global Economy: Bringing Domestic Institutions Back In*. Cambridge: Cambridge University Press, pp. 145–168.

Wong, J. (2011) *Betting on Biotech: Innovation and the Limits of Asia's Developmental State*. Ithaca: Cornell University Press.

9
Conclusion: Regenerative Medicine – A New Paradigm?

Andrew Webster

Introduction

The preceding chapters have shown how the boundaries of the field of regenerative medicine (RM) are far from stable, and how this is true whether one focuses on its local or global contexts. Indeed, it is the interaction between these two contexts that creates much of the tension, uncertainty, and activity in the field, illustrated, for example, by stem cell tourism, by the move towards international standards in research paralleled by competing local conventions and regulation, and by the competition between innovation models in the 'West' and China and Japan. Both Chapters 2 and 3 have shown that these dynamics create a complex mix of corporate activity allied to clinical trials across different global regions, notably in the United States, in Germany, the United Kingdom, and France in Europe, and in China, South Korea, and Japan in East Asia. The tissue economy being built is highly uncertain and will require a wide range of codifying, standardising, and authorising (via regulatory approval) moves to be made. As we have argued in different ways across the chapters, these socio-technical challenges merely reflect the ways in which RM is a disruptive technology, a form of innovation that is poorly aligned with existing regulatory infrastructures, clinical practices, and commercial markets. In order to grow, as in any other emergent field, such as nanotechnology or synthetic biology (Calvert, 2012), RM networks need to be built and form, in Latour's (2005) terms, new 'assemblages' that bring together diverse material, technological, and political entities to form a *socially* robust field called regenerative medicine. The notion of a 'field' has to be seen in this sociological sense rather than more conventionally seen as being purely a biological domain of inquiry. As was noted in Chapter 1, the boundary of the field is best understood as being enacted (Mol and Law, 2004) or performed rather than developed according to a logical, linear sequence of steps. Bioscience sets those priorities that identify problems – again of a material, technological, and political nature – that are more likely to be answerable while leaving many other questions unanswered, shelved for another time.

At the same time, we have of course sought to provide some analytical boundaries that define our primary area of inquiry and while this posed some measurement problems – as Graham Lewis discussed in Chapter 2 – we have adopted two broad positions throughout the book. First, that there are a number of centrally important biomaterial features of this area that mark off its specific nature, namely, the process of deploying tissue in novel ways to regenerate body function. This is clearly new and helped set the broad parameters of our discussions throughout. But, secondly and equally importantly, what these features mean in regard to putting this into practice, in assembling the entities needed to make this work, is a thoroughly social, emergent, and contested process that is anchored in differing expectations, hype, and competing normative positions about matters of 'life'. In combining these two dimensions – the materiality and the sociality – of RM we can ask how far it might be said to provide a new model or paradigm for medicine that is distinctive from those currently prevailing.

Those working within RM tend to adopt two rather different positions on its revolutionary nature, some arguing that, while there have been significant steps taken in the past decade or so, especially via the isolation of embryonic stem cell lines in the later 90s and the generation of induced pluripotent stem (iPS) cells more recently, the field builds on much longer standing work going back 30 years or so when bone marrow transplants were used in the treatment of various – especially blood – disorders. As was noted in Chapter 1 and elsewhere in the book, others such as Mason (2007) have argued that this earlier period should be characterised as 'RM 1.0', while the more recent developments 'RM 2.0', similar to the shift seen in social media analysis from Web 1.0 to Web 2.0. This second argument places most emphasis on the transformative impact of stem cell therapies rather than the more broadly defined boundary of the field marked out by the first approach.

Even if we recognise the important developments related to stem cell science, it is still arguable whether this is tantamount to or of similar nature to the transformations seen in the digital universe described as Web 2.0. A defining feature of Web 2.0 is the way in which the content and design of the web is generated by non-experts via a whole series of social media including crowdsourcing and the hybrid role of being a 'prosumer' creating what Hardey (2008) called a 'public culture of research' that contrasts with formal, expert-based science.

What we would argue here in contrast to the view that there is an RM 2.0 emerging is that while Web 2.0 itself might be associated with the field through, for example, Internet-based patient activism pressing for more funding in the area, or through the global web-based movements associated with stem cell tourism, we do not (yet?) see *the field itself* sharing the characteristics of Web 2.0: people are not developing or co-creating their own clinical trials; RM is yet to call for a move towards crowdsourcing of data; and we do not see the emergence of lay RM on Facebook. While we need, therefore, to be cautious about claims for the arrival of RM 2.0., might it be said to offer a new form of medicine that contrasts with the principal domains found today?

What might the latter be? Typically, conventional medicine is divided into devices (such as pacemakers), medicinal products (notably drugs), and surgical procedures. These work with/in the body either through prosthetic enabling of the body, remedying or masking a pathological cellular process or, as in the case of surgery, by direct intervention to correct, remove, or insert body functionality (the latter, for example, through tissue or organ transplantation). RM is distinctive in seeking to promote the body's own restorative powers through either transplanted cells or triggering endogenous repair of cells within the body. To the extent that this is possible RM does appear to be a very different form of medicine, one which must work with the material properties of *live* tissue. This is very different from whole organs used for transplantation since the live tissue has to be derived from cell sources (such as embryonic stem cell lines), and manipulated in such a way that it will become a stable cell type, such as muscle, nerve, skin, heart tissues, differentiated to perform one particular function in the body. Growing on such cells is biologically difficult. One reason for this is the inherent variability in stem cells as they are grown on to form batches of cells. While this poses problems for both the producer of cells and the regulation of them, the problem also creates difficulties for the scaling up of lines, as noted in Chapter 1. Overall, then, it does appear to be the case that RM is a significant, if not radical, shift in medicine's 'ways of knowing' and intervening in (Pickstone, 2000) the body.

However, as we have seen throughout the book, the regulatory oversight of the field has meant that, as a *social object of regulation*, RM bleeds into *existing* domains of medicine and therapy (as in tissue engineering, the use of cell-infused scaffolds, etc.) such that its distinctiveness is much less clear. Not only might it be allied with a device (as in a scaffold), it might be defined *as a device* (as in the construction and insertion of an artificial pancreas). It was the uncertainty surrounding where it sat on the regulatory spectrum that led the European Medicines Agency and European Commission to define a new category of the 'Advanced Therapy Medicinal Product' (with its three subdivisions of 'tissue engineered products', 'advanced somatic cell therapy products', and 'gene therapy products') as we saw earlier in the book. Moreover, the very material on which RM depends – such as stem cells – can and has been defined as the *equivalent to a drug*. In July 2012, for example, the US Food and Drug Administration (FDA) declared that stem cell lines should be regarded as drugs where they involve the manipulation of cells prior to implantation. Since drugs are required to be licensed for use by the FDA, this meant the FDA could close down what it regarded as unsafe stem cell clinics in Colorado, that, until then, had been injecting autologous cells into patients, regarding this as a routine medical practice and so not subject to FDA approval. While the deployment of drugs-based legislation in this way is used to police the field, it can also be used to enable its commercial development inasmuch as some firms see the 'cells-as-drugs' model as the only way in which RM will be possible to scale up with regulatory approval and to be both clinically widely available and commercially profitable. The irony here is that those who see RM as radical and transformative are at the same time envisaging a clinical and business model that is highly conventional, one that is typically found in the pharmaceutical sector.

From what we have suggested thus far, it would seem that the most judicious answer to the question 'is RM paradigm-changing?' is to say that it is breaking new clinical boundaries in terms of its biological/material goals and processes but it is socially located in a regulatory and commercial context that means that changes will be incremental and move at different paces on various fronts (e.g., autologous vs. allogeneic) and that early adoption is most likely to be within hospitals as part of the existing 'hidden innovation system'. At the same time, more broadly, RM has posed some significant problems for the regulatory domain, and is, in the expression of Haddad et al., in Chapter 4 a 'novel experimental site of contemporary bio-politics'. In part this explains why regulatory oversight, which tends to be risk averse, will be incremental, as steps are taken to deal with one uncertainty then another, but in doing so opening up new problems that need dealing with, and has knock-on effects for other areas of regulation, pointing to an ongoing and 'inevitable incompleteness of any attempts of governing' (ibid.). This is then as much a question about how far-existing regulatory and innovation paradigms can encompass and so enable the development of RM, and how far they can do so without radical change, as it is whether RM signals a paradigm shift in medicine itself.

The social shaping and framing of RM is most evident when we consider the diverse ways in which it has developed at the global level and how relations between the global and local develop. We have seen, especially in the later chapters of the book, how the field has been shaped by different innovation and regulatory processes within and between countries. In China we saw competing regulatory models framing stem cells as either a drug (akin to the FDA's line) or as a medical technology; how the Chinese have sought to adopt international (i.e., in effect Western originating) standards of governance yet accommodate without any formal oversight very different treatment and evidence-based regimes as provided by Beike, which, though based in China, operates at a global level via the Web. Chapter 5 examined the ways in which the procurement of oocytes (egg harvesting) for somatic cell nuclear transfer (SCNT) research was dependent on links between the regenerative and reproductive bioeconomies, but also showed how the risks and burdens that women face cannot be understood via an abstracted globalised or 'universal' process, but is highly context-specific and varying by locale. Similarly, while there are moves towards a global bioethics – a universal set of principles that can underpin research and practice in the RM field – Sandor and Varju's chapter shows how this butts up against local legal regimes (especially as expressed in property rights) which are more, or less, able to accommodate them. In the encounter between ethics and law, they make the crucial point that the meaning or status of RM bio-objects as enshrined in various ethical principles may be redefined by the precepts of law which vary by country (recall the contrast between the restrictive regime of Germany to the more permissive of the United Kingdom). In a countervailing process, Salter describes how global competition in the RM field has led to the emergence of a more sophisticated and complex pattern of governance, operating at local, regional, and transnational levels, and serving to enable, in the round, the growth of the bioeconomy.

In general, then, the global dynamics of RM echo globalising processes seen elsewhere in the world economy: growing interdependency, mobilisation of human, scientific, and raw material resources, standardisation and codification of knowledge and in particular scientific research and results, paralleled at the same time by a more complex and diverse set of circumstances at the local level through which these global processes are mediated.

In light of the remarks above that begin to draw together the conclusions of this book, what are the main themes and issues that form the basis of a social science critique of RM?

Work within science and technology studies (STS) points to the importance of the materiality of objects, here of bio-objects, and the ways in which *novel forms of life* challenge conventional cultural, scientific, and institutional orderings and classifications, including the meaning of 'life' itself (Holmberg et al., 2011): what, for example, iPS cells 'actually are'. Jasanoff has in a similar vein spoken of 'ontological surgery' (2011) to evoke both the biological, visceral manipulation of life *and* the ways in which this reshapes the meaning of human identity itself. RM, in particular through its work on developing embryonic stem cell lines, does this and does so in such a way as to pose new problems for those performing ethical and regulatory roles associated with the oversight and policing of science. At the same time, STS points in contrast to the *obduracy* of materiality, to the difficulty of orchestrating the biological to perform in the way you want. Combining these two complementary notions, the critical issue is how novel but undisciplined life gets to be put to work in clinical settings. This raises practical questions, for example, about patient safety (see Chapter 3 above), and the need for new forms of risk assessment that go beyond the conventional approaches seen, for example, in drug delivery (i.e., 'pharmacovigilance'). One key issue and one which will remain for the foreseeable future is how to manage a patient's immune response and ensure that there are limited adverse effects from stem cell implants through tracking (e.g., via genetic tags) how cell differentiation is working in vivo. It is also important to note that a viable regulatory environment will also recognise a difference between stem cell therapies that follow a cell replacement model of application where cells and their descendant populations are intended to make a long-term contribution to an individual's body tissue, and stem cell therapies, mainly using what are called mesenchymal stem cells, which are intended to have a short-lived stimulatory effect on the body's innate regeneration systems at the site of application.

These issues are recognised by those working in the RM field: it is the role of a social science critique to show, through tracing the debates over the novelty and obduracy of material bio-objects, how the uncertainties and risks so created must inform regulatory oversight, one that is, in a phrase coined many years ago by the STS/science policy scholar David Collingridge (1980), undertaken 'under conditions of ignorance'. Haddad et al. in this book have shown very clearly how 'unruly' RM objects can be, and hence, how decision-making structures and processes are often unable to anticipate the risks and impact of the technology. Instead, such processes should, informed by Collingridge's precept, be subject to open and recursive forms of vigilance rather than closure. Recent analysis within STS of the clinical trial in RM (Webster et al., 2011) exemplifies how the social science/bioscience interface can be mutually productive in opening up debate about a key stage in the development of new therapies.

A second area for social science critique focuses on the RM bioeconomy and how, as we have noted earlier in the book, the generative potential of biological life becomes intertwined with that of (bio)capital. It is clear from what we have argued that there is no single innovation path but a number of them currently being pursued by small biotech firms and the global pharma companies. Innovation analysis looks for the appearance of niche areas for development (Schot & Geels, 2008) or for the gradual onset of 'path dependencies' (Rycroft and Kash, 2002) whereby a new technological regime begins to close off options and lock in to one preferred development path. This is partly because of the interplay of positive feedback factors both internal and external to a field – standardisation often plays a key role in this regard (not just in RM, think too of standard rail gauges). At this point in time, niche analysis is more appropriate than path dependency in understanding the RM field since it is yet to see the alignment of biotechnological, institutional, and organisational structures that would act as selection mechanisms that would lock into specific development paths across the field as a whole. This point about alignment ties in with a wider theme within STS that emphasises the importance of seeing how the technological and organisational aspects of a new field of science co-evolve, each dependent on the other to build and stabilise the area. So, for example, there is a recursive relation embedded in the question 'what are the data requirements that will meet the regulatory requirements for clinical trials?' More generally, we can argue that gradual moves towards the Europeanisation of the field through the Advanced Therapy Medicinal Products (ATMP) and harmonisation of clinical trials procedures are not only building stability but building a regulatory version of Europe itself.

A third area to which social science can contribute relates to the last point just touched on: regulation and governance. Recent work by Tait (2011) and Tait et al. (2011) has suggested that 'path-breaking technology' does not need 'path-breaking regulation', that is to say overly restrictive regulation and governance of a field that could kill it off in its early stages of development. They argue for 'smart regulation' that is fit for purpose. This is related to whether innovation is radical or disruptive rather than incremental, making the case for less onerous regulation where incremental steps are being taken in a sector: as Tait argues,

> An innovation that challenges a sector's internal [research and development – R&D] model and at the same time its regulatory and market environments is much more likely to be seriously disruptive or path-breaking than one which affects only one of these areas.
>
> (2011, p. 2)

Tait argues that for smaller tissue engineering companies, stem cells are likely to be incremental rather than disruptive in their impact on the innovation path taken and the products that appear, whereas they are likely to be much more disruptive if adopted by big pharma through changing the basis on which future drugs are developed. This is a sensible approach to adopt, but, as Tait and colleagues themselves note, the regulation/innovation relation is complex, and as we have argued throughout this book, the degree to which innovation is *seen to be* incremental or radical varies considerably within and between regulatory domains and national arenas. In other words, the attributions of 'incremental' versus 'radical' are in important ways socially constructed. Moreover, regulatory provisions, as we saw in the discussion of the Oviedo Convention in Chapter 6, acted not as a brake but more of a device that could be either worked around or paradoxically deployed to assist in stem cell research, despite the Convention's restrictive clauses. Thus, apparently restrictive legislation need not act as such at all, whereas in other settings one might want to welcome restrictive legislation if it helps secure patient safety, as the FDA's policy towards private stem cell clinics in the United States aims to do.

How far restrictive legislation is in fact to be found is in any case open to debate since Salter in Chapter 8 shows how governance has become a vehicle through which states and regions (such as the European Union (EU), US, India, and China) can compete in the RM field. To the extent this is happening it would seem that there are many who know how to play the 'smart regulation' game already, even though, as Salter observes, they still have to resolve the 'conundrum of the relationship between old and new forms of innovation governance' (see Chapter 8 above). Emergent economies are currently building their own governance models for oversight of the RM field and it is clear that these struggle to find some consistent framework that brings together the interests and priorities of Salter's three forms of governance – social, market, and scientific.

Our final comment relates to how a social science understanding of the field might provide insight into the likely therapeutic applications of RM in the future. We argued at various points in the book that we are likely to see clinical development at least in the medium term in the area of autologous cell therapies rather than allogeneic ones, since for allogeneic therapies to emerge they will require innovation in other areas, especially in terms of the governance of assessment and the evaluation of cell therapies relative to existing forms of treatment. There will need to be considerable input made from a specific domain of social science, namely, health economics, to develop more nuanced models of cost-effectiveness and outcome measures. There are also a number of issues that are especially challenging for allogeneic therapies, namely, managing the problems of immune-suppression of patients' immune response, sourcing donor cells, large-scale culture systems, the purification, storage, and transport of cell lines to clinics, and the clinical skills and knowledge needed for safe implantation. Within the European and US contexts it appears likely therefore that autologous stem cell therapies developed within academic and/or clinical settings are the preferred route to treatment. In part this is reflected in the fact that investment by large pharma and venture capital firms is going into the areas that are not only clinically but also ethically less challenging. We expect that we are likely to see the appearance of new institutional and governance arrangements to underpin the development of in vitro fertilisation (IVF)-style clinics for autologous cell therapy and perhaps the growth of hospital-based treatment, which is much more the model already adopted in Japan and to some extent in China and India. Both would require much greater engagement with RM by public health care systems which we suspect would be of especial interest to smaller firms with limited independent resources. This would be more than the 'hospital exemption' route (which we discussed in Chapter 3), which was in itself never intended as a route through which products would be developed for markets – patenting and charging do not enter the frame here. In the meantime, we are likely to see pockets of stem cell treatment activity in the smaller private stem cell clinics, such as Regenerative Sciences in Colorado under challenge from the FDA. In the longer term, large pharmaceutical investment will depend on the 'cells-as-drugs' model being seen to be successful, and that will depend ironically on the work that smaller firms such as Reneuron in the United Kingdom and ACT in the United States play in breaking new ground with regulators and in clinical trials.

Conclusion

We began this chapter by asking whether RM provides a new paradigm for medicine. We hope that we have shown in this chapter and the book more generally that the way this is often framed is much too restrictive, focusing almost entirely on the biological novelty and therapeutic promise of the field. Instead, the notion of paradigm shift has to be one which involves a combination of bioscience, institutional, and cultural changes. It may well be that in the much longer term the more socially significant paradigm change will have to occur in the regulatory context in order to cope with the unruly objects and therapeutic uncertainties that cell therapies may generate. We spoke in the first chapter about the two processes of 'enclosure' and 'instability', and how the play of different interests in the RM field help secure the first, and others the second. In mapping out the global dynamics of RM we have tried to show how it occupies a very uneven landscape.

In some ways, what we might try to think through here is what Oudshoorn (2011) has, in a very different setting via her work on telecare/telemedicine, called a 'techno-geography' of the field. In Oudshoorn's work the term is used to refer to the ways in which place, care, and technology are configured in the delivery of telecare, with a particular emphasis on understanding the importance of place and the relationship between formal and informal care (institutional and 'extitutional' care). The field of RM has a range of clinical settings in which therapies are and will become available, involving different clinical and business models, differing ways in which risk and responsibility for treatment, clinical practice, and subsequent care is distributed (Akrich and Passveer, 2002), and different spatial patterns and levels of regulation and control of the field. We have used the term boundary and landscape throughout our discussion and suggest that the concept of 'technogeography' can be used heuristically to draw our attention to the ways in which a field works to reconfigure the relationship between people, place, and technology, and to ask how, and how far, this is and is likely to be true of RM. It may well be the case that some of the developments are regarded as simple extensions of existing biomedical practices while others more ostensibly 'radical'. It is likely that the dynamics of enclosure and instability will determine the outcomes we will see and thereby the spatial character of the field itself.

What we are likely to see, and indeed are seeing at present, is a quite uneven development of the field on different scientific, commercial, and clinical fronts, as the play and counterplay of the dynamics of enclosure and of instability take their course. The geographical and institutional patterns and the practices of different actors found therein are therefore both complex and requiring of sustained social science analysis in order to map the likely regional and global developments over the next decade and more. We hope that the chapters in this book have contributed towards such an analysis.

References

Akrich, M. and B. Passveer (2002) Multiplying obstetrics: techniques of surveillance and forms of coordination, *Theoretical Medicine*, 21: 63–83.

Calvert, J. (2012) Ownership and sharing in synthetic biology: a 'diverse ecology' of the open and the proprietary?, *BioSocieties*, 7(2): 169–187.

Collingridge, D. (1980) *The Social Control of Technology*. London: Pinter

Hardey, M. (2008) Public health and web 2.0, *Perspectives in Public Health*, 128(4): 181–189.

Holmberg, T., N. Schwennesen and A. Webster (2011) Bio-objects and the bio-objectification process, *CMJ*, 52(6): 740–742.

Jasanoff, S. (2011) *Reframing Rights: Bioconstitutionalism in the Genetic Age*. Boston: MIT Press.

Latour, B. (2005) *Reassembling the Social: An Introduction to Actor-Network-Theory*. Oxford: Oxford University Press.

Mason, C. (2007) Regenerative Medicine 2.0, *Future Medicine*, 2(1): 11–18.

Mol, A.-M. and J. Law (2004) Embodied action, enacted bodies: the example of hypoglycaemia, *Body and Society*, 10(2–3): 43–62.

Oudshoorn, N. (2011) *Telecare Technologies and the Transformations of Healthcare*. Basingstoke: Palgrave Macmillan.

Pickstone. J. (2000) *Ways of Knowing: A New History of Science, Technology and Medicine*. Manchester: Manchester University Press.

Rycroft, R. W. and D. E. Kash (2002) Path dependence in the innovation of complex technologies, *Technology Analysis and Strategic Management*, 14(1): 21–35.

Schot, J. and F. W. Geels (2008) Strategic niche management and sustainable innovation journeys: theory, findings, research agenda, and policy. *Technology Analysis and Strategic Management*, 20(5): 537–554.

Tait, J. (2011) *Multinational Company Innovation Strategies*, Innogen Policy Brief. Available at http://www.genomicsnetwork.ac.uk/media/AGLS1

Tait, J., J. Chataway and D. Wield (2011) *The Case for Smart Regulation*, Innogen Policy Brief. Available at http://www.genomicsnetwork.ac.uk/media/ AGLS2

Webster, A., C. Haddad and C. Waldby (2011) Experimental heterogeneity and standardisation: Stem cell products and the clinical trial process, *BioSocieties*, 6: 401–419.

Index

Printed in the United States
by Baker & Taylor Publisher Services